Guide to
JCT Standard
Building Contract 2016

JCT Standard Building Contract
With Quantities (SBC/Q)

JCT Standard Building Contract
With Approximate Quantities (SBC/AQ)

JCT Standard Building Contract
Without Quantities (SBC/XQ)

RIBA  **Publishing**

Sarah Lupton

Guide to JCT Standard Building Contract 2016

© Sarah Lupton, 2017

Published by RIBA Publishing, part of RIBA Enterprises Ltd
The Old Post Office, St Nicholas Street, Newcastle upon Tyne, NE1 1RH

ISBN 978 1 85946 640 7

British Library Cataloguing-in-Publication Data
A catalogue record for this book is available from the British Library.

Commissioning Editor: Elizabeth Webster
Project Editor: Alasdair Deas
Designed and typeset by Academic + Technical, Bristol, UK
Printed and bound by Page Bros, Norwich, UK
Cover design: Kneath Associates
Cover image: Shutterstock: www.shutterstock.com

RIBA Publishing is part of RIBA Enterprises Ltd.

www.ribaenterprises.com

Foreword

Long accepted as an industry standard, the JCT Standard Building Contract has a tried and tested track record, lending a reassuring familiarity to those whose job it is to administer it. The new edition, SBC16, continues the familiar logical layout and clear drafting style, which makes it easy to adopt and easy to run. Changes in working practices and case law are all reflected in the new versions of the contract.

Sarah Lupton's *Guide to JCT Standard Building Contract 2016*, which follows on from her excellent *Guide to SBC11*, is a straightforward and comprehensive analysis of the new form in the light of today's legal and practice landscape. Not only does she point out the important new changes, particularly to those provisions concerned with payment and insurance, but she reflects on recent cases which serve as valuable reminders. The contract's provisions, procedures and supplementary conditions are spelled out knowledgeably and are organised by theme. The hard-pressed practitioner will be particularly pleased to see the useful indexes, and will doubtless come to depend on being able to dip quickly into the book for specific help during the course of a job.

I would also thoroughly recommend the book to both architecture and other construction students on the threshold of undertaking their professional examinations. The comprehensive up-to-date coverage clearly and succinctly exposes the legal ramifications of the contract. Sarah Lupton's rare combination of being a legally-trained architect who also runs the MA in Professional Studies at the Welsh School of Architecture makes this book the ideal student companion.

Neil Gower, Solicitor
Chief Executive, The Joint Contracts Tribunal
May 2017

About the author

Professor Sarah Lupton MA, DipArch, LLM, RIBA, CArb is a partner in Lupton Stellakis and directs the Master of Design Administration and the Diploma in Professional Practice at the Welsh School of Architecture. She is dual qualified as an architect and as a lawyer. She lectures widely on subjects relating to construction law and is the author of many books including this series on JCT contracts, the *Guide to the RIBA Domestic and Concise Building Contracts*, *Which Contract?* and the 5th edition of *Cornes and Lupton's Design Liability in the Construction Industry*. She contributes regularly to the International Construction Law Review and acts as an arbitrator, adjudicator and expert witness in construction disputes. Sarah is also chair of the CIC's Liability Panel and the CIC Liability Champion.

Contents

About this Guide

The JCT Standard Building Contract is the longest established contract published by the JCT, and its origins can be traced back to the nineteenth century. As such, it has a long track record of adjustment in response to legislation, court judgments and evolving industry practices and is considered a benchmark against which other standard forms and bespoke contracts are frequently compared. It is widely used, and even where its use is not proposed, a thorough knowledge of the Standard Building Contract will help practitioners to understand and appraise any contract they encounter.

The fourth edition of this guide covers the latest version of the JCT Standard Building Contract, published in 2016 (previous editions covered the 1998, 2005 and 2011 forms). Some significant changes have been introduced into SBC16. The interim and final payment procedures have been redrafted to consolidate and simplify the clauses, and the result is significantly shorter and clearer than the previous version. The procedures and time limits for notification and assessment of loss and expense have also been tightened up and clarified. In addition, the contract now incorporates and updates provisions from the JCT Public Sector Supplement relating to Fair Payment principles, transparency and BIM. Provisions for performance bonds and parent company guarantees have been added, and arrangements to secure provision of third party rights/warranties from sub-contractors have been revised.

The Guide is intended primarily for consultants, such as architects or surveyors, who are advising clients on contracts and who may be acting as contract administrators. However, it is frequently used by others in the construction field, for example contractors, sub-contractors and clients, and is popular with students. It will also be of use to lawyers who need an introduction to the form or a quick reminder of its key features.

The Guide retains the structure and style of previous editions. It does not assume any prior knowledge of either SBC16 or any previous editions of the form. The subject is covered by topic, rather than a clause-by-clause commentary. Brief introductions to legal concepts are included where necessary to understand the form's provisions. Short case summaries are included for court cases cited in the text, to help practitioners understand the context of the legal point being made. However, the Guide focuses on the form itself, its practical application and the roles of and decisions to be made by the parties and the contract administrator.

1 About SBC16

Key features

1.1 The JCT Standard Building Contract 2016 (SBC16) is the most recent edition of the longstanding JCT Standard Form of Building Contract. It is intended for use in traditional procurement, i.e. the design is largely complete at tender stage, but the form also makes provision for limited contractor design (the contractor's designed portion). The current edition is published in three versions: With Quantities, Without Quantities and With Approximate Quantities. The differences between them relate to the documents on which the contract sum is based, and the calculation of sums due to the contractor. Unless otherwise indicated, this Guide makes reference throughout to the With Quantities version (SBC/Q), with attention drawn to other versions where the differences are significant.

1.2 All three versions of the form are intended for use by both private clients and local authorities. The contract particulars indicate that certain provisions (e.g. cl 4.7, provisions for advance payment and an associated bond) do not apply where the employer is a local or public authority. Special provisions are included which are required by or essential to its use by local authorities and public bodies, as discussed below at paragraph 1.17.

1.3 In larger construction programmes a traditional contract may be linked to an umbrella framework agreement, or used in connection with single-project partnering. Although originally developed in the private sector, this approach is also often used in the public sector, and SBC16 includes provisions that reflect this context. There is reference to a framework agreement (seventh recital), which would be used alongside the form; and as an alternative, in cases where there is no partnering agreement, several of the optional supplemental provisions in Schedule 8, such as collaborative working and key performance indicators, reflect a partnering ethos and could be used as a basis for single-project partnering.

1.4 As a traditional form, SBC16 is relatively simple in its overall structure. The contractor must carry out the work shown in the contract documents for the sum entered in the contract particulars, and within an agreed time period. Use of the form requires the appointment of an 'Architect/Contract Administrator', a quantity surveyor, a principal designer and a principal contractor, all of whom are named in the articles. There are provisions for varying the work, and adjusting the contract sum and the completion date on the occurrence of certain events, all provisions shared with most other traditional standard forms of contract.

1.5 The contractor's primary obligation is to carry out the work shown or described in the contract documents, and the contractor takes overall responsibility for ensuring that the standards set out in the contract documents are achieved. In general terms, the form assumes that all work is designed by the employer's design team, and that the contractor will be supplied with all information necessary to carry out the works.

1.6   The contractor may, however, be required to carry out the design of a 'Contractor's Designed Portion'. The provisions relating to this requirement in many ways reflect those of the JCT Design and Build Contract. The part of the works to be designed is identified in the recitals, which also refer to the employer's requirements and the contractor's proposals for the contractor's designed portion, and to the contractor's analysis of the portion of the contract sum relating to the contractor's designed portion, termed the 'CDP Analysis'. The requirements are sent out with the tender documents and the proposals returned with the tender, together with the analysis. The form includes a procedure for the submission of the related design information by the contractor for comment by the contract administrator, who retains responsibility for integrating the contractor's designed portion with the rest of the design. The contractor is required to carry insurance to cover its design liability.

1.7   The contract administrator has a significant role under the contract, which includes the issuing of certificates and the power to order variations to the works. At some points the contract administrator is acting as the employer's agent, and at others as an independent administrator. A court would assume that the parties have contracted on the basis that the contract administrator will act fairly at all times in applying the terms of the contract and particularly so when acting as an independent administrator. This duty of fairness, however, does not place the contract administrator in the same position as an arbitrator, in that the contract administrator is not immune from being sued.

1.8   The form makes provision for three different methods of sub-contracting: to a domestic sub-contractor selected by the contractor and approved by the contract administrator; to a sub-contractor chosen from a list of three named in the contract documents; or to a 'named specialist', who may be identified in the contract documents or in an instruction relating to a provisional sum. With the contract administrator's approval the contractor's design obligations may be sub-contracted to a domestic sub-contractor. There are now two JCT sub-contracts specifically for use with SBC16: the Standard Building Sub-Contract, and the Standard Building Sub-Contract with Sub-Contractor's Design. The contractor is required to sub-let on these terms where appropriate, and there is a requirement for specific terms to be incorporated in any sub-contract.

1.9   The contract requires that the contractor commences work on an agreed 'Date of Possession', and completes the works by an agreed 'Date for Completion'. The contractor is required to produce a master programme, although the form does not set out any particular requirements for the programme, or a sanction for its non-production. There are provisions which allow for the date for completion to be adjusted on the occurrence of specified events, and for the contractor to pay damages in the event of non-completion.

1.10   SBC16 allows for phased working, in that it is possible to divide the works to be carried out into sections, and set separate start and completion dates in relation to each section. In general the conditions relating to timing operate independently with respect to each section. For example, the contractor must notify the contract administrator of delays to any section, and there is provision for fixing new completion dates for each section as appropriate. A separate practical completion certificate is required for each section, and a separate certificate of making good, but only one final certificate. There are implications for non-completion, liquidated damages and retention.

1.11   Both the With Quantities and Without Quantities versions are lump sum contracts. In other words, all the work 'shown, described or referred to' in the contract documents must be carried out for the contract sum 'or such other sum as shall become payable under this Contract'. The amount of work, which is covered by the contract sum, should be described in exact terms in the contract documents. In the With Quantities version the work is described in drawings and in a bill of quantities, whereas in the Without Quantities version the description is in drawings and a specification or schedules of work. Generally, if the description is inaccurate, any resulting addition to the cost is borne by the employer. If the contractor has made an error in pricing, however, then any shortfall will be borne by the contractor. The contract administrator has wide powers to order variations to the works if required, and the contractor has a corresponding right to be paid any additional costs that arise from such variations.

1.12   JCT With Approximate Quantities is a remeasurement contract, where only approximate quantities are given for all of the work to be carried out, and the contract assumes that all work will be measured prior to certification. This version would normally be used where it may be difficult or impossible to measure the majority of the work accurately in advance, for example in a contract for refurbishment or repair following a fire or other damage. An approximate quantity may also be given for identified items when using SBC16 With Quantities. All versions allow for the use of provisional sums where it is impossible to specify or describe the work accurately in advance.

1.13   Payment to the contractor is made upon the issue of contract administrator's certificates at predetermined intervals. In general terms, the certificates will reflect the amount of work that has been properly completed up to the point of valuation in accordance with the terms of the contract. The provisions regarding payment, including the requirements for notices and the contractor's rights in the event of non-payment, reflect those required under the Housing Grants, Construction and Regeneration Act (HGCRA) 1996 Part II, as amended by the Local Democracy, Economic Development and Construction Act (LDEDCA) 2009 Part 8 (see Table 1.1).

1.14   In addition to the core provisions concerning quality, design, programme and payment, SBC16 also contains detailed provisions covering matters such as injury, damage and insurance, third party rights and warranties, default and termination, and dispute resolution, all of which are discussed in detail in this Guide.

Table 1.1 Provisions required under the HGCRA 1996 as amended by the LDEDCA 2009

HGCRA 1996: section	SBC16: clause	Provision concerning
108	Article 7/9.2	Right to refer disputes to adjudication
108A	n.a.	
109	4.8	Stage payments
110(1)	4.11	Dates for payment
110(1A–1D)	n.a.	
110A	4.9.1	Payment notices
110B	4.10.1/4.10.2	Payee notices
111	4.11.5	Notice of intention to withhold payment
112	4.13	Right to suspend performance

1.15   The form follows the normal JCT format of agreement, recitals, articles, contract particulars, attestation and conditions. The seventh recital makes reference to a framework agreement which may supplement the provisions of the contract. The form also includes nine 'Supplemental Provisions' under Schedule 8. These optional provisions are referred to in the eighth recital, and are incorporated by indicating whether or not they are to apply in the contract particulars (note that the supplemental provisions generally apply unless stated otherwise, with the exception of Supplemental Provisions 7 and 8, which apply only where the employer is a local or public authority, and Supplemental Provision 9 (Named Specialists), which applies only if selected (see footnote [14]).

1.16   The form contains an advance payment bond (Schedule 6: Part 1), a bond in respect of payment for off-site goods and materials (Schedule 6: Part 2) and a retention bond (Schedule 6: Part 3). The JCT collateral warranties from the contractor to a funder (CWa/F) and purchaser or tenant (CWa/P&T), which are published separately, may be used with SBC16, as may the JCT collateral warranties from a sub-contractor to a funder (SCWa/F), purchaser or tenant (SCWa/P&T) and employer (SCWa/E).

1.17   SBC16 has many features required for public sector procurement, including Fair Payment provisions, transparency and the use of BIM (building information modelling). These were originally published as a supplement to the JCT contracts, but have now been incorporated into the form with some further amendments. The Fair Payment provisions arise from the stated aims of the government in *Construction 2025*, which include equitable financial arrangements and certainty of payment throughout the supply chain. The aims are reflected in initiatives such as the Construction Supply Chain Payment Charter 2014, the HGCRA 1996 (as amended), the Late Payment of Commercial Debts Regulations 2013 and the Fair Payment Charter, as well as regulation 113 of the Public Contracts Regulations 2015. These require that the final date for payment should be 'no later than the end of a period of 30 days from the date on which the relevant invoice is regarded as valid and undisputed' (regulation 113(2)(a)), and that similar provisions are included in sub-contracts and sub-subcontracts. Under the charter, the value of work and materials supplied by all three tiers is to be assessed as at the same date. Adopting SBC16 together with the appropriate JCT sub-contracts will ensure that the government requirements are met. The contract also includes provisions (Supplemental Provision 7) relevant to any employer that is subject to the Freedom of Information Act 2000 (which would include local authorities). This provides that the parties accept that the contract is not confidential, except for material that may be 'exempt' and which the employer has the discretion to determine. The Public Contracts Regulations 2015 also deal with corrupt practices and bribery, and breach of the statutory requirements is a ground for termination under clauses 8.6 and 8.11.3 of SBC16. Furthermore, under Supplemental Provision 8 the contractor must include similar provisions in any sub-contract.

1.18   The JCT publishes a guide for use with SBC16 (the *Standard Building Contract Guide*, SBC/G). This gives general guidance on the scope of the clauses and the changes since SBC11. It also includes a model form for the rights particulars and a list of related JCT publications, at appendices A and B.

Deciding on SBC16

1.19   The JCT Standard Form of Building Contract has been the first choice of form for many contemplating a traditional procurement route. Long accepted as an industry standard, it has a tried and tested track record that gives users the reassurance that there should be little to surprise them, and much guidance is available on its use. However, there are some aspects that should be given careful consideration before deciding to proceed with this form.

1.20   Compared with some other traditional forms, such as the JCT Major Project Construction Contract (MP), this form places more risk on the employer. One of the primary functions of any construction contract is to allocate risk between the parties. In many ways it is meaningless to describe any particular contract as being 'fair' or 'unfair' in that, as long as the allocation of risk reflects what the parties intended, they enter into it as a commercial decision and the price agreed would reflect that balance. The more risk the employer is prepared to accept, the lower the price is likely to be. A good example from SBC16 is the operation of the fluctuations clauses, which allow the contract sum to be adjusted to take account of changes in market prices. If this provision is not used then the contractor must anticipate the risk of price changes, and this may result in a relatively high tender.

1.21   A further example is that the responsibility for providing the contractor with all information reasonably necessary to construct the project lies with the employer, except for work forming part of the contractor's designed portion. Even here, the contractor is not responsible for checking the employer's requirements, and is entitled to proceed with its design on the assumption that all information provided is accurate. Any delay in providing information, or in commenting on information provided by the contractor, is the employer's risk. The contractor is also entitled to claim additional time upon the occurrence of various neutral events, such as exceptionally adverse weather, which is not the case in some other standard form contracts. It is therefore important that anyone advising an employer on the possible selection of SBC16 as the contract form to be used on a project understands how the balance of risk between the parties is affected.

1.22   SBC16 can admittedly be criticised for being lengthy and complex in its procedural rules. The construction of large projects is, however, a complex process often involving many risks, therefore in order for parties to decide in advance what should be the outcome of various eventualities the form is necessarily detailed and long. By comparison it should perhaps be pointed out that there are far longer forms of building contract in regular use today; for example, the Engineering Advancement Association of Japan (ENAA) Model Form of Contract runs to five volumes! An alternative approach, and that adopted by MP, is to provide the parties with a set of core clauses, and require them to draft additional terms to suit their particular needs, but this method requires expert advice, and is likely to produce a document at least as long as SBC16, or otherwise the parties will be left 'in the dark' as to their relative positions in certain circumstances. Administrators of large and complex projects must be prepared to undertake the sophisticated procedures and levels of complexity that attend major forms of contract.

1.23   JCT Ltd has had to respond to changes in working practices, to take account of new legislation and to act where the judiciary has indicated that parts of the form are unclear, or has sometimes placed an interpretation on the form that it was never intended to have. As a result of this the provisions of SBC16 are balanced and finely honed. To carry out ad hoc amendments can produce an imbalance and bring unexpected consequences.

1.24   SBC16 contains some features that are unique among JCT forms for traditional procurement, including the procedures for 'variation quotations' and 'acceleration quotations', the provisions for 'third party rights' and the option of a retention bond (see Table 1.2). The introduction of the 'named specialists' facility in 2012 will have increased the appeal, as this had been lacking since the removal of the nominated sub-contractor provisions in 2005 (although it should be noted that the named specialists provisions differ significantly from those for nominated subcontractors). In situations where these features are an advantage, SBC16 will no doubt remain a popular choice.

Table 1.2 Comparison of JCT contract provisions						
	MP	SBC	ICD	IC	MWD	MW
Contractor design	yes	yes	yes		yes	
'Contractor's Design Submission Procedure'	yes	yes	yes			
Possession by sections		yes	yes	yes		
Completion by sections	yes	yes	yes	yes		
Deferment of possession		yes	yes	yes		
Information release schedule		yes	yes	yes		
Partial possession	yes	yes	yes	yes		
Employer's representative	yes	yes				
Clerk of works		yes	yes	yes		
Listed sub-contractors		yes				
Named sub-contractors/specialists	yes	yes	yes	yes		
Pre-appointed consultants	yes					
Advance payment, bond		yes	yes	yes		
Activity schedule		yes	yes	yes		
Application by contractor	yes	yes	yes	yes		
Payment for off-site materials, bond		yes	yes	yes		
Retention bond		yes				
Performance bond or guarantee		yes	yes	yes		
Variation quotations	yes	yes				
Acceleration quotations	yes	yes				
Interest on late payment	yes	yes	yes	yes	yes	yes
Professional indemnity insurance	yes	yes	yes			
Third party rights	yes	yes				
Code of Practice for tests		yes				
Fluctuations options		1	1	1	1	1

Note: MP = Major Project Construction Contract
ICD = Intermediate Building Contract with Contractor's Design
IC = Intermediate Building Contract
MWD = Minor Works Building Contract with Contractor's Design
MW = Minor Works Building Contract

Changes in the 2016 edition

1.25   Prior to the 2016 edition, the JCT had published two sets of adjustments to the 2011 edition: the 'Named Specialist Update' (February 2012), a new optional method for introducing client-selected sub-contractors, and Amendment 1 (March 2015), relating to the CDM Regulations 2015. The JCT had also published a Public Sector Supplement in 2011, for use by local or public authorities. These changes are now all incorporated in the 2016 edition.

1.26   The key 2016 changes are set out in Table 1.3 and can be summarised as follows:

- incorporation and updating of provisions from the JCT Public Sector Supplement relating to Fair Payment principles, transparency and BIM;
- amendments relating to the CDM Regulations 2015;
- reference made to various provisions of the Public Contracts Regulations 2015;
- changes in respect of payment, designed to reflect fair payment principles and to simplify and consolidate the payment provisions;
- payment is monthly, including during the rectification period;
- removal of Fluctuations Options B and C – these are now published separately;
- the inclusion of performance bonds and parent company guarantees;
- tightening up of the arrangements to secure provision of third party rights/warranties from sub-contractors;
- simplification and rationalisation of drafting in many areas; for example, calculation of amounts due and provisions for rectification following damage covered by insurance.

Table 1.3 Key changes

SBC16 clause	New/revised	Key changes
Contract particulars	new	New Supplemental Provisions 7 (transparency), 8 (Public Contracts Regulations) and 9 (named specialists) added
Contract particulars	new	Reference to BIM Protocol added
Contract particulars	revised	Dates of interim valuation replace 'due dates'
Contract particulars	new	Entries in respect of performance bonds and guarantees
Contract particulars	revised	Third party rights and warranty information to be set out in separate 'Rights Particulars'
1.1	revised	'BIM Protocol', 'C.1 Replacement Schedule', 'Consultants', 'Design Submission Procedure', 'Employer Rights', 'Existing Structures', 'Interim Valuation Date', 'Local or Public Authority', 'Measurement Rules', 'Named Specialist', 'Named Specialist Work', 'Payment Application', 'Payment Certificate', 'Payment Notice', 'PC Regulations', 'Post-Named Specialist Work', 'Pre-Named Specialist Work', 'Principal Designer', 'Rights Particulars' and 'Works Insurance Policy' added to defined terms

Table 1.3 Key changes – Continued

SBC16 clause	New/revised	Key changes
1.4	revised	References to documents to include information in a format set out in any BIM protocol
1.9	revised	Effect of conclusiveness in relation to dispute resolution clarified
1.11	new	All consents and approvals not to be unreasonably delayed or withheld
2.29.2.3 and 2.29.14	new	New relevant events relating to named specialists
2.41.3	new	Provision for licence to be assignable
3.2	revised	Provision for 'person-in-charge' deleted and 'Site Manager' added
3.7	revised	Reference to named specialists added
3.9.2.5	new	Sub-contracts to provide for granting of third party rights or execution of warranties
3.9.3	new	Sub-contracts to provide for supply of information and grant of licences in relation to the BIM protocol
3.23	revised	CDM Regulations 2015 clarified
4.3	revised	General adjustments simplified
4.8 and 4.10	revised	Due dates, applications and payment notices relating to interim valuation dates clarified
4.11.1	revised	Final payment to be 14 days from due date (not 28 days as in SBC11)
4.14 and 4.15	revised	'Sum due' calculation clarified
4.20	revised	Loss and expense application procedure simplified
6.2, 6.3	revised	Contractor's liability for damage due to its negligence clarified
6.13, 6.14	revised	Procedures for making good after damage redrafted and simplified; included in main contract text, not in schedules
7.3	new	New provisions for performance bonds and guarantees
7.4	revised	Parties to set out details of recipients of third party rights or warranties in a separate document (rights particulars)
7E	revised	Warranties/third party rights from sub-contractors redrafted and clarified
8.11.3		Right to terminate in relation to the Public Contracts Regulations 2015 added
Schedule 3	revised	Drafting of insurance options clarified
Schedule 3	new	Provision for a 'C.1 Replacement Schedule' added
Schedule 7	revised	Fluctuations Options B and C removed (published separately)
Schedule 8	new	Supplemental Provisions 7 and 8 for transparency and the Public Contracts Regulations 2015, respectively, added
Schedule 8	new	Supplemental Provision 9 for named specialists added

2 The contract documents

2.1 A contract entered into on the basis of SBC16 will comprise an extensive package of documents, the majority of which will have been issued to the contractor at tender stage. Documents are central to the success of every building operation, and in traditional procurement in particular the contractor depends on full and accurate information being provided in adequate time and to a predetermined pattern. Generally, firm and full information at tender stage reduces the risk of cost increases and programme alterations later in the contract.

2.2 Ideally, the formal contract documents should be executed before the project commences on site. Normally, a contract is formed if there is a clear acceptance of a firm offer.[1] The contract, once executed, will supersede any conflicting provisions in the accepted tender and will apply retrospectively (*Tameside Metropolitan BC* v *Barlow Securities*).

> ***Tameside Metropolitan Borough Council* v *Barlow Securities Group Services Limited* [2001] BLR 113**
>
> Under JCT63 Local Authorities, Barlow Securities was contracted to build 106 houses for Tameside. A revised tender was submitted in September 1982 and work started in October 1982. By the time the contract was executed, 80 per cent of building work had been completed, and two certificates of practical completion were issued relating to seven of the houses in December 1983 and January 1994. Practical completion of the last houses was certified in October 1984. The retention was released under an interim certificate in October 1987. Barlow Securities did not submit any final account, although at a meeting in 1988 the final account was discussed. Defects appeared in 1995, and Tameside issued a writ on 9 February 1996. It was agreed between the parties that a binding agreement had been reached before work had started, and the only difference between the agreement and the executed contract was that the contract was under seal. It was found that there was no clear and unequivocal representation by Tameside that it would not rely on its rights in respect of defects. Time began to run in respect of the defects from the dates of practical completion; the first seven houses were therefore time barred. Tameside was not prevented from bringing the claim by failure to issue a final certificate.

2.3 When using SBC16, the primary document is of course the printed form itself, which comprises not only articles and conditions but also various schedules, which include 'third party rights', forms of bonds and fluctuations provisions (see Table 2.1). SBC16 also makes reference to various other documents. Some of these are termed 'Contract Documents' (see Table 2.2) whereas others are referred to at various places in the form, for example the recitals refer to an 'Activity Schedule' and an 'Information Release Schedule', and the articles refer to the Construction Industry Model Arbitration Rules – all significant documents. Although not termed 'Contract Documents', many of these may form part of

---

[1] See, for example, Stephen Furst and Vivien Ramsey (eds), *Keating on Construction Contracts*, 10th edn (London: Sweet & Maxwell, 2016) or Peter Aeberli, *Focus on Construction Contract Formation* (London: RIBA Publishing, 2003).

Table 2.1 Layout of the form

Agreement:

- Recitals
- Articles
- Contract Particulars
- Attestation

Conditions:

- Section 1: Definitions and Interpretation
- Section 2: Carrying out the Works
- Section 3: Control of the Works
- Section 4: Payment
- Section 5: Variations
- Section 6: Injury, Damage and Insurance
- Section 7: Assignment, Performance Bonds and Guarantees, Third Party Rights and Collateral Warranties
- Section 8: Termination
- Section 9: Settlement of Disputes

Schedules:

- Schedule 1: Design Submission Procedure
- Schedule 2: Variation and Acceleration Quotation Procedures
- Schedule 3: Insurance Options
- Schedule 4: Code of Practice
- Schedule 5: Third Party Rights
- Schedule 6: Forms of Bonds
- Schedule 7: JCT Fluctuation Option A
- Schedule 8: Supplemental Provisions

the contract between the parties. Indeed any document to which clear reference is made in the 'Contract Documents' will form part of the binding agreement between the parties.

2.4   The documents used will, to a certain extent, depend upon the version of SBC16 that is selected. In addition, if the works are to include a contractor's designed portion, this will affect the documents to be used. Table 2.2 indicates some of the possible combinations of documents that may make up the contract package.

'Contract Documents'

2.5   SBC16 With Quantities defines 'Contract Documents' as 'the Contract Drawings, the Contract Bills, the Agreement and these Conditions, together with (where applicable) the Employer's Requirements, the Contractor's Proposals and the CDP Analysis and (where applicable) the BIM Protocol' (cl 1.1). The agreement and conditions are, of course, found in the form. SBC16 With Approximate Quantities also includes contract bills within the definition of 'Contract Documents', although in this case all the quantities will be approximate. SBC16 Without Quantities defines contract documents as including '(where Pricing Option A applies) the Priced Document or (where Pricing Option B applies) the Specification'. The former is where the contractor priced the specification or work schedules (the 'Priced Documents'). The latter is where the contractor has stated a lump sum only and is required

Table 2.2 Documents	With Quantities	Without Quantities	With Approximate Quantities: Pricing Option A	With Approximate Quantities: Pricing Option B
Drawings	CD	CD	CD	CD
Bills	CD	CD		
Priced specification or works schedules			CD	
CDP analysis or schedule of rates				CD
Specification				CD
Rule 11b schedule (fluctuations Option C only)			CD	CD
Priced activity schedule (optional)	R		R	R
Information release schedule (optional)	R	R	R	R
Employer's requirements (if CDP used)	CD	CD	CD	CD
Contractor's proposals (if CDP used)	CD	CD	CD	CD
CDP analysis (if CDP used)	CD	CD	CD	CD
BIM protocol	cl 1.1, 1.4	cl 1.1, 1.4	cl 1.1, 1.4	cl 1.1, 1.4
Construction phase plan	cl 2.1	cl 2.1	cl 2.1	cl 2.1
Contractor's master programme	cl 2.9.1.2	cl 2.9.1.2	cl 2.9.1.2	cl 2.9.1.2
Contractor's design documents	cl 2.9.4	cl 2.9.4	cl 2.9.4	cl 2.9.4
Advance payment bond	cl 4.7	cl 4.7	cl 4.7	cl 4.7
'Listed Items' and related bond	cl 4.16	cl 4.16	cl 4.16	cl 4.17
Retention bond	cl 4.18	cl 4.18	cl 4.18	cl 4.18
'C.1 Replacement Schedule'	cl 6.7	cl 6.7	cl 6.7	cl 6.7
Performance bond	cl 7.3	cl 7.3	cl 7.3	cl 7.3
'Rights Particulars' (third party rights/warranties)	cl 7.4	cl 7.4	cl 7.4	cl 7.4

Note: CD = termed a 'Contract Document'; R = referred to in the recitals; CDP = contractor's designed portion; cl = referred to in the conditions.

in addition to supply either a contract sum analysis or a schedule of rates on which the lump sum is based, which is referred to as 'the Priced Document', but is not defined as a contract document in clause 1.1. In both cases, where Fluctuations Option C applies, the schedule required by rule 11b of the JCT Formula Rules will also be a contract document.

2.6 The agreement comprises the recitals, articles, contract particulars and the attestation. The recitals, articles and contract particulars must be completed very carefully. The attestation must be signed by both parties and witnessed. This forms the heart of the agreement whereby the contractor undertakes to 'carry out and complete the Works in accordance with the Contract Documents' (Article 1), and in return the employer undertakes to pay the contractor the contract sum as adjusted in accordance with the conditions (Article 2).

Contract drawings

2.7   The 'Contract Drawings' are listed under the third recital (or the second recital in the Without Quantities version). These should all be identified precisely, including revision numbers, etc. The list may be annexed if long, but if so the list must be clearly identified, and the drawings should be 'signed or initialled by or on behalf of each Party' (second recital). Note that in SBC16 there is no reference to the party responsible for preparing the drawings.

Contract bills

2.8   The 'Contract Bills', unless otherwise stated, must be prepared in accordance with the 'Measurement Rules' which are the RICS *New Rules of Measurement – Detailed Measurement for Building Works* (NRM2) unless otherwise stated (cl 2.13.1). If materials or goods are to be paid for prior to delivery on site, a list of these must be annexed to the bills (cl 1.1, definition of 'Listed Items'). Bills of quantities are normally prepared by the quantity surveyor, who is named in Article 4 (this is the case even if the contract administrator is to act as quantity surveyor, see footnote [9]). The bills are based on detailed drawings and a specification prepared by the contract administrator. The use of bills does not reduce the responsibility of the contract administrator for the preparation of that information. For good practice in preparation and co-ordination of specification, drawings and bills of quantities see current relevant publications by the Construction Project Information Committee (CPIC). The contract requires that the parties sign both the drawings and the bills (second and third recitals).

2.9   In practice, even where bills of quantities are used, a specification often forms part of the documentation, either bound in as a section of the bills, perhaps as part of the preamble, or as a separate document referred to in the bills. The CPIC publications recommend that the specification becomes the core document in terms of defining quality, and that the drawings and the bills refer to clauses in the specification. If this system is used, the specification will be an essential part of the package and it is suggested that it should be signed with the other 'Contract Documents'.

2.10   The third recital (Alternative A) of the Without Quantities edition refers to 'the Specification or Works Schedules'. One of these should be deleted in the contract particulars as appropriate, though there is no reason why both a specification and schedules should not be used in the contract package. The contract does not define what form the schedules should take, but the documents could be arranged by work sections, by trades or, as is frequently used in refurbishment, on a 'room-by-room' basis. In all cases it is likely to be clearer if the detailed specification information is kept in a separate document referred to by the schedules. If this is done then it is suggested that, as the contractor would normally be asked to price the schedules, the reference to the priced specification is deleted in the contract particulars. The specification should then be bound into or identified and referred to in the schedules, and signed with the other contract documents.

Employer's requirements

2.11   The 'Employer's Requirements' are referred to in the tenth recital as 'documents showing and describing or otherwise stating his requirements for the design and construction of the Contractor's Designed Portion', and the form assumes that these have been sent to the contractor at tender stage.

2.12   The contract does not stipulate any format for the employer's requirements. In broad terms, the documents will set out the employer's requirements for the contractor's designed portion of the works. The requirements should be prepared carefully and on the assumption that there will be no changes to the requirements once the contract is let, for although the contract contains provisions whereby a variation can be instructed, such variations may result in additional costs to the employer, and are subject to the consent of the contractor.

2.13   The employer's requirements could be in a very summary format, for example simply giving a brief description of the relevant part or system, referring to drawings indicating its location and co-ordinating dimensions. It is likely, though, that they will be more detailed than that and include a detailed specification, in either prescriptive or performance terms, or in all probability involving a mixture of the two.[2] They could also include schematic layouts or outline designs of the relevant part. In essence, the employer's requirements act as a brief.[3] Where descriptive or performance specifications are included, these should be accurate. Use of the phrases 'to be to the contract administrator's approval' or 'to be approved' should be avoided at all costs (see paragraph 3.23).

2.14   One of the most important inclusions is to stipulate in exactly what form the contractor's proposals should be submitted, and what they should include. This is essential in order for the employer to make a clear assessment of the submitted tenders. The amount and level of detail of the information will depend upon the scale of the contractor's designed portion, and its relationship with the rest of the design. Where the contractor's designed portion forms a significant element in the project, full information may be needed in order to integrate this element with other elements of the design.

2.15   It is also very important that the requirements should specify the drawings and other design information (the 'Design Documents') to be submitted by the contractor following acceptance of tender, and a programme for their submission. The purpose of this is to control the scope, format and timing of the submission of design documents for review. For example, it should protect the contract administrator from being overwhelmed by design documents at an inconvenient time, or from being presented with design documents to review for key elements in isolation from information on other related aspects of the design. It is likely that the programme will be the subject of negotiation at tender, as it is important that any programme in the requirements will also meet the contractor's needs in terms of developing the design at a rate which will support its intended construction programme. It would also be wise to set out the information required to be submitted at practical completion, such as 'As-built Drawings', otherwise the contractor's obligation is to provide such information 'as the employer may reasonably require' (cl 2.40).

---

[2]   Guidance on the preparation and use of performance specifications can be found in the *JCT Guide to the Use of Performance Specifications* (London: RIBA Publishing, 2001).

[3]   Advice on briefing is outside the scope of this Guide, but reference could be made to Paul Fletcher and Hilary Satchwell, *Briefing: A Practical Guide to RIBA Plan of Work 2013 Stages 7, 0 and 1* (London: RIBA Publishing, 2015).

Contractor's proposals

2.16   The contractor's proposals should be in the format and contain the information stipulated in the employer's requirements. These may request that various documents are provided, including drawings, specifications, schedules, programmes, method statements, etc.

2.17   The contractor should raise matters relating to the contract data where decisions are outstanding from the employer, so that these can be resolved. The proposals should indicate clearly any areas of conflict in the requirements, and any instances where the contractor has found it necessary to amend or amplify the brief. The contract does not allow for the inclusion of provisional sums in the proposals, only in the requirements (cl 3.16), so if the contractor wishes to cover any part of the proposals with a provisional sum then it should inform the employer so that the requirements can be amended.

CDP analysis

2.18   The contract does not prescribe a format for the CDP analysis. It would therefore be sensible to set out what format would be acceptable in the employer's requirements. (It would not be unreasonable, in cases where the contractor's designed portion forms a significant part of the works, for the contractor to be asked to prepare a full bill of quantities, although this would be unlikely on smaller projects.) The contract requires that the document is used for assessing the value of employer-instructed variations to the contractor's designed portion (cl 5.8.2). It does not require that the document is used to assess the value of work carried out, etc., to be included in periodic payments, but it would normally be used by the contractor to prepare applications for payment, and by the employer in checking such applications.

BIM and other protocols

2.19   If building information modelling (BIM) is to be used on the project, it will be important that the parties agree many matters to do with how the model will be prepared and managed, such as format, communication methods, timing, the detecting and resolving of problems, copyright, use following completion, etc. The contract allows for the use of a 'BIM Protocol', and the parties should adopt a standard form protocol or prepare a bespoke one for the project. At the time of writing, the only standard form available is the *Building Information Model (BIM) Protocol* published by the Construction Industry Council (CIC, 2013, available free from the CIC website). The protocol to be used should be identified in the contract particulars (cl 1.1). Clause 1.4.6 states that 'references to documents shall, where there is a BIM Protocol or other protocol relating to the supply of documents or other information, be deemed to include information in a form or medium conforming to that protocol'. (Note that cl 1.1 refers only to a BIM protocol; if some other protocol is needed then this would require minor amendment.)

Other documents

Activity schedule

2.20   The second recital (third recital in the Without Quantities version, and not in the With Approximate Quantities version) refers to a priced 'Activity Schedule', a provision that should be deleted if not required. The schedule is prepared and priced by the contractor and provided prior to the contract being executed. An example of a priced activity schedule is included in the *Standard Building Sub-Contract Guide* (SBCSub/G), and it is very similar to a schedule of work (see Appendix A to this Guide). Each activity is priced, and the sum of those prices will be the contract sum, with certain exclusions, such as provisional sums and approximate quantities (see footnote [3] to the second recital, or footnote [4] to the third recital in the Without Quantities version). If it is included, it is used to ascertain the value of work properly executed for certification purposes (cl 4.14.1.1).

2.21   An activity schedule is unlikely to be useful where full 'Work Schedules' have been prepared by the design team, unless they have been prepared on a different format to that suggested for the activity schedule, for example on a 'room-by-room' basis. It may be useful alongside bills of quantities if it saves time, and therefore reduces consultant's fees, in relation to interim certificates, but may result in a loss of accuracy of valuations. It is likely to be most helpful where the tender package consists of drawings and specification alone.

Information release schedule

2.22   The 'Information Release Schedule', referred to in the fifth recital, is an optional provision (the recital is deleted if it is not required). The schedule should state 'the information the Architect/Contract Administrator will release and the time of that release'. If used, the schedule should be prepared by the contract administrator and sent out with the tender documents. The schedule does not need to list all the information that will be provided but should, for example, list key drawings.

2.23   The schedule will of course make it clear to the contractor in advance when information will be provided, and will therefore enable the contractor to programme the work more effectively, and possibly reduce the number of potential arguments that may arise regarding delays. A significant implication for the contract administrator is that if any information listed is provided later than the stipulated date then this will be a relevant event in relation to an extension of time. It should also be noted that if there is any adjustment to the completion date then adjustments to the schedule might have to be negotiated between the parties.

Health and safety documents

2.24   The employer and contractor are required to comply with the Construction (Design and Management) (CDM) Regulations 2015 (cl 3.23). A key element of the Regulations is the employer's duty to appoint a principal designer and a principal contractor (regulation 5). On most projects using SBC16 the contract administrator will be the principal designer, and the contractor the principal contractor.

2.25   The construction phase plan is not a contract document under SBC16, and the recitals make no mention of it having been prepared and given to the contractor at the time of tender. However, the employer (and the principal designer, if not the contract administrator) must provide the contractor with pre-construction information (regulations 4(4) and 12(3)), which should be sent out with the tender documents. Where the contractor is the principal contractor, it must ensure that the construction phase plan is prepared before setting up the construction site (regulation 12(1)); compliance with this is required under clause 3.23.2. To avoid uncertainty, it is advisable to require that this document be submitted by the contractor well in advance of the start of work on site. Following commencement, the contractor must ensure that the plan is reviewed and updated on a regular basis (regulation 12(4)).

2.26   Under the Regulations the health and safety file is principally a matter for the principal designer, who will compile it (regulation 12(5) and (6)), but there is a requirement on the contractor to provide information for the file (regulation 12(7)). Under SBC16, clause 2.30 requires the contractor to have complied with all its CDM duties with respect to the supply of documents and information before a certificate of practical completion is issued.

Bonds

2.27   SBC16 includes three forms of bond: (1) an advance payment bond, (2) a bond in respect of payment for off-site materials and/or goods and (3) a bond in lieu of retention. Where required, the contractor must arrange bonds, and as all of these are optional it must be made clear to the contractor at tender stage if any will be required. An advance payment bond is normally required where an advance payment is to be made to the contractor under clause 4.7 (note that this is not an option where the employer is a local authority). A bond in respect of payment for off-site materials and/or goods is required where it has been agreed that certain materials or goods will be paid for in advance of them being brought on site – the so-termed 'Listed Items' (cl 4.16). A retention bond is an alternative form of security to the more traditional use of a retention deduction (cl 4.18). Terms for all of the bonds have been agreed between the British Bankers' Association and JCT Ltd, and are included in the form under Schedule 6. SBC16 also refers to a performance bond (cl 7.3), but in this case does not include a form. If a performance bond is required, or any other type of bond, then the terms must be available to the contractor before the contract is entered into.

Sub-contract documents

2.28   JCT Ltd publishes two versions of a standard form for use with domestic sub-contracts (one for use where a design obligation is to be sub-contracted, and one for use where it is not) and although there is no requirement under SBC16 that the main contractor should use this form, there are restrictions on the terms that may be agreed. These are set out in clause 3.9 of SBC16, which requires, for example, that particular conditions relating to ownership of unfixed goods and materials, and the right to interest on unpaid amounts properly due to the sub-contractor, are included in all domestic sub-contracts. The sub-contract should also, of course, comply with the requirements of the Housing Grants, Construction and Regeneration Act (HGCRA) 1996 Part II (as amended by the Local Democracy, Economic Development and Construction Act 2009).

Use of documents

Interpretation, definitions

2.29   Section 1 of the form sets out some rules governing the interpretation of the conditions. Clause 1.1 is a schedule of definitions of terms that are used throughout the contract. Some further and more detailed definitions are embodied in the text of clauses, for example 'All Risks Insurance' and 'Joint Names Policy' are defined in clause 6.8. Clause 1.4, defines what is meant by any reference to a 'person', or to a statute, and also contains a gender bias clause. Section 1 also includes items first introduced by Amendment 18 to JCT80. Those regarding notices and periods of time were required by the HGCRA 1996 Part II and re-state its requirements relating to the calculation of periods of days and the serving of notices (cl 1.5 and 1.7). Clause 1.7 allows the parties to agree that certain communications, to be identified in the contract particulars, may be made electronically. The parties may also agree an exact format for the electronic communications. Otherwise, all communications are to be in writing.

Priority of contract documents

2.30   Clause 1.3 states 'The Agreement and these Conditions are to be read as a whole. Nothing contained in any other Contract Document or any Framework Agreement, irrespective of their terms, shall override or modify the Agreement or these Conditions'. If this clause were not included, the position under common law would be the reverse; in other words, anything that had been specifically agreed and included in a document would normally override any standard provisions in a printed form.

2.31   If the parties wish to agree to any special terms that differ in any way from the printed conditions, then the amendments will need to be made to the actual form, which usually involves the insertion of one or more additional articles. If necessary, due to lack of space, these amendments could refer to the special terms, which could be appended to the form or included in the bills of quantities. Amending standard forms is unwise without expert advice as the consequential effects are difficult to predict. Deleting clause 1.3 could be particularly unwise as it might have unintended effects on other parts of the contract. (If significant changes are needed, consider the use of another form, perhaps the Major Project Construction Contract, which embraces individual tailoring and has no equivalent clause.)

Inconsistencies, errors or omissions

2.32   The contractor is under an obligation to point out any discrepancy or divergence within or between the contract documents, including the CDP documents, and/or any further instructions, documents or drawings issued by the contract administrator (cl 2.15). The obligation appears to be limited to those discrepancies that the contractor has discovered. There is no obligation for the contractor to search for discrepancies, although the general obligation to use reasonable skill and care would suggest some degree of observance could be expected. Any notice should be issued immediately upon discovery and should include 'appropriate details' of the error or discrepancy. If the contractor fails to point out any discrepancies that it notices, or should have noticed, and work has to be re-done as a result, then the contractor may lose any right to extra payment, extension of time, and loss and/or expense. The contract administrator's obligation to issue an instruction under clause 2.15 appears to be limited to instances where the contractor has found a discrepancy. However, the contract administrator's general obligation to provide necessary information would extend to correcting any errors and discrepancies.

Errors in the contract bills

2.33   The contract requires that any error in the contract bills shall be corrected (cl 2.14.1). It does not say by whom but this would be the responsibility of the employer and normally carried out by the quantity surveyor. The correction is treated as if it were a variation required by a contract administrator's instruction (cl 2.14.3), and though there is no express requirement to issue such instruction, the correction should be confirmed in writing.

The contract bills and contract drawings

2.34   The contract does not specifically deal with the situation where there is a divergence between the information shown in the contract bills and that set out on the contract drawings. However, as clause 4.1 states that 'The quality and quantity of the work included in the Contract Sum shall be that set out in the Contract Bills', in the case of conflict the contract bills will normally take precedence. If some other result is preferred then the contract administrator will need to issue an instruction, which will constitute a variation.

Employer's requirements and contractor's proposals

2.35   Where there is an error in the contractor's proposals or the CDP analysis, this is corrected, but is not to result in any addition to the contract sum (cl 2.14.4). The contractor must inform the contract administrator of its proposed amendment to deal with any discrepancy within or between the contractor's proposals and/or any other contractor design document. The contractor is obliged to accept the contract administrator's decision and comply at no cost to the employer (cl 2.16.1). If the contract administrator failed to reach a decision within a reasonable time, this could be grounds for an extension of time and loss and/or expense.

2.36   Where there is a discrepancy within the employer's requirements, or a discrepancy between the requirements and any variation, the contract states that if the contractor's proposals deal with the discrepancy then they will prevail (cl 2.16.2). The discrepancy between the requirements and any variation refers to inadvertent problems resulting from the effect of a variation, rather than intended alterations to the particular part of the requirements at which the variation was aimed. If the employer decides it does not like the solution in the contractor's proposals and would prefer some other solution, this would have to be instructed as a variation.

2.37   If the contractor's proposals do not deal with the discrepancy, the contractor is required to inform the employer of its proposed amendment for dealing with it, and the contract administrator must either agree or decide on alternative measures and, in either case, notify the contractor in writing (cl 2.16.2). The acceptance or notification is to be 'treated as a Variation', which would result in it being valued under clause 5.2, and in it constituting grounds for an extension of time under clause 2.29.1, for loss and/or expense under clause 4.22.1, and for termination under 8.9.2.1, in the unlikely event that it causes a suspension (see paragraph 5.43). If there was undue delay by the contract administrator in reaching a decision, then this would also be grounds for a claim.

2.38   A footnote to the twelfth recital gives advice on how to deal with a divergence between the employer's requirements and the contractor's proposals that is identified before the contract is executed; in that case the issue should be resolved and the documents amended to reflect this prior to execution. The contract does not deal with the situation where such a divergence is discovered after the contract is formed.

2.39   This recital is somewhat problematic from the point of view of the employer. Although not a condition of the contract it nevertheless, in the absence of any contrary provision, appears to give precedence to the contractor's proposals. The guidance notes to an earlier edition of the design and build form state that the intention of this recital is that it should be without prejudice to the contractor's liability in respect of design, and that, for example, if the employer's requirements included a performance specification for a heating system, and the employer subsequently accepted the contractor's design proposals for the system, the employer would not be precluded from alleging breach of contract. This may have been the intention of the drafters but the position is far from clear.

2.40   If there were a discrepancy between the CDP documents, a court would endeavour to determine, from an objective standpoint, what were the true intentions of the parties, and might well decide that the employer should be deemed to have accepted the version set out in the contractor's proposals, at least in so far as the divergence would have been revealed by a reasonably thorough examination. As the employer may prefer its own requirements to take precedence, this recital is frequently altered to that effect in practice.

Divergences from statutory requirements

2.41   The contractor and the contract administrator are both required to notify each other of any discrepancy or divergence between any of the clause 2.15 documents, or any instruction requiring a variation, and any statutory requirement as defined under clause 1.1 (cl 2.17.1). Where the discrepancy relates to the employer's requirements, the contractor's proposals or other contractor design documents, the contractor must inform the contract administrator of its proposed amendment to deal with the discrepancy. In all cases the contract administrator must issue instructions to deal with the problem. Where the divergence relates to the employer's requirements, the contractor's proposals or other contractor design documents, the contract states that the contractor must comply at no extra cost to the employer, unless the divergence results from a change in statutory requirements since the base date (cl 2.17.2.1). In all other cases the instruction is treated as a variation (cl 2.17.2). The effect of this clause is that the costs will be borne by the contractor in situations where the divergence is between the employer's requirements and statute, as well as between the contractor's proposals and statute.

Custody and control of documents

2.42   The contract drawings and contract bills remain in the custody of the employer, and must be available for inspection at all reasonable times (cl 2.8.1). The contract administrator should retain a copy for reference throughout the life of the contract. The contractor must be provided with one certified copy and two further copies, unless a BIM protocol is used which requires otherwise (cl 2.8.2).

2.43   The documents provided must not be used for any purpose other than the works, and the details of the rates or prices are not to be divulged (cl 2.8.4). The contractor must keep on site at all reasonable times one copy of all the contract documents, the contractor's design documents and all other documents listed in clauses 2.9 to 2.12, which includes unpriced bills of quantities, the master programme and further schedules and information issued by the contract administrator (cl 2.8.3). At practical completion the contractor must provide copies of drawings and information relating to the contractor's designed portion as stipulated in the contract documents, or as the employer may reasonably require (cl 2.40).

Assignment and third party rights

Assignment

2.44   The right of a subsequent purchaser to bring an action against the builder of their property, with whom they have had no contractual relationship, could be of considerable value. The employer in a construction contract might therefore wish to assign this right to such other person who may acquire an interest in the property.

2.45   A contractual right can be regarded as a personal right of property, and in property law it is classified as a 'chose in action'. Choses in action can be assigned under the Law of Property Act 1925, provided the requirements of section 136 of the Act are followed. It is important to note that it is only contractual rights that can be assigned, termed 'the benefit' of a contract, and not obligations. So if, for example, A enters into a contract with B whereby A agrees to carry out some building work and B agrees to pay A £100 for the work, A can assign the right to claim the £100 to C but not the obligation to carry out the work. The right to pursue a debt or claim is assignable to C without B's consent, provided B is notified as required by section 136. The obligation to carry out the work, however, could only be transferred to C with the agreement of all three parties (often termed 'novation').

2.46   SBC16 contains express provisions which limit the scope for assigning contractual rights. Clause 7.1 states that neither the employer nor the contractor may 'assign this Contract or any rights thereunder' without the written consent of the other. Assignment without consent of the other party is grounds for termination (cl 8.4.1.4 and 8.9.1.3). There is one exception, however, to the prohibition on assignment: if clause 7.2 is stated to apply in the contract particulars then the employer may assign some limited rights to a party to whom it has transferred a freehold or leasehold interest in the premises comprising the works. Among other limitations the rights can only be assigned after practical completion. The clause does not provide a general right to assign the benefit, but the right to bring proceedings in the name of the employer to enforce terms of the contract made for the benefit of the employer. It is thought that this would limit the assignee to claiming at most losses suffered by the employer as a result of any breach by the contractor, and would not extend to further losses suffered by itself.

Third party rights/warranties

2.47   SBC16 offers two options for the granting of rights to bring a claim to persons who are not a party to the contract, either through the use of the 'third party rights' provisions included in the form, or through the use of separately published standard form warranties.

2.48   The 'third party rights' provisions make use of the facility introduced by the Contracts (Rights of Third Parties) Act 1999. Until this Act came into force, it was a rule of English law that only the two parties to a contract had the right to bring an action to enforce its terms (termed 'privity of contract'). However, it is often the case in construction projects that other parties may wish to be in a position to be able to take action, should one or other of the parties default on their obligations. A future owner of the property may, for example, wish to be able to claim against the contractor should it later transpire that the project was not built according to the contract. Under the rule of privity, the future owner would be a third party, and would not be able to bring a claim. In response to this, 'collateral warranties' were developed which allowed for third parties to pursue claims for breaches of a contract. Examples of such warranties would be between contractor and owner, contractor and funder, and also between consultants and owners/funders.

2.49   The Contracts (Rights of Third Parties) Act has changed the fundamental rules of law relating to privity, in that it entitles third parties to enforce a right under a contract, where the term in question was to provide a benefit to that third party. The third party could be specifically named, or could be of an identified class of people. The effect of this Act is therefore to open the door to the possibility of claims being brought by a range of persons, in some cases persons that the parties to the contract may never have considered.

2.50   The Act, however, allows for parties to agree that their contract will not be subject to its provisions, and many standard forms adopt this course in order to limit the parties' liability. SBC16 takes this approach and under clause 1.6 states:

> Other than such rights of any Purchasers, Tenants and/or Funders as take effect pursuant to clauses 7A and/or 7B, nothing in this Contract confers or is intended to confer any right to enforce any of its terms on any person who is not a party to it.

2.51   The contract therefore by this clause 'contracts out' of any effects of the Act. (In the light of the above, it is important to note that the effects of deleting or amending this clause would be significant.) It then allows the parties to define exactly which third parties will have rights with respect to the contract (in a separate document called the 'Rights Particulars', clauses 7A.1 and 7B.1) and what those rights will be (Schedule 5).

2.52   Schedule 5 sets out 'Third Party Rights for Purchasers or Tenants' (Part 1) and 'Third Party Rights for a Funder' (Part 2). The contractor warrants (in relation to the tenant) that it has carried out the works in accordance with the contract (with effect from practical completion), and (in relation to the funder) that it has complied with and will continue to comply with the contract. This allows both the purchaser/tenant and the funder to bring an action in respect of breaches of contract by the contractor.

2.53   There are some things to note about this system. In the case of purchasers and tenants, the contractor's liability extends to the reasonable costs of repair, renewal or reinstatement, but does not include other losses unless so stated in the contract particulars (Schedule 5, Part 1:1.1), in which case the liability will be limited to a stated maximum amount. The contractor's liability is also limited by a net contribution clause (Schedule 5, Part 1:1.3). Under the arrangement the contractor is entitled to rely on any term in the contract should any action be brought against it by a third party (Schedule 5, Part 1:1.4). Where there is a contractor's designed portion, the contractor is required to provide evidence of its professional indemnity insurance, if requested, to any person possessing rights under the Third Party Rights Schedule (Schedule 5, Part 1:5). The rights may be assigned by the purchaser or tenant without the contractor's consent to another person, and by that person to a further person, but beyond this no further assignment is permitted (Schedule 5, Part 1:6).

2.54   In the case of the funder, except for the inclusion of a net contribution clause (Schedule 5, Part 2:1.1), no limit is placed upon the extent of the contractor's liability. As above, the contractor is entitled to rely on any term in the contract should any action be brought by the funder (Schedule 5, Part 2:1.2), and the rights may be assigned by the funder without the contractor's consent to another person, and by that person to a further person, but beyond this no further assignment is permitted (Schedule 5, Part 2:10). The Schedule also sets out various 'stepping in' rights which may be exercised by the funder in the event that it terminates its finance agreement with the employer.

2.55   Under the alternative system of 'collateral warranties' the contractor has actually to enter into a warranty separately with each beneficiary. The beneficiaries are identified in the 'Rights Particulars', and the warranties are identified in clauses 7C and 7D as the JCT standard forms of warranty to purchaser/tenant and funder (CWa/P&T and CWa/F). The warranty forms comprise identical terms to the third party rights set out in Schedule 5.

Procedure with respect to third party rights and warranties

2.56   Where third party rights are to be used, the relevant details must be set out carefully in the 'Rights Particulars' section, which is identified in the contract particulars. It is important to identify the funder/purchaser/tenant because if none is identified the rights/warranties shall not be required. It is not necessary, however, to identify a specific organisation; the description could simply be of a class of persons, e.g. 'all first purchasers' or 'the lead bank providing finance for the project'.

2.57   The third party rights take effect from the date of receipt by the contractor of the employer's notice to that effect; in the case of a purchaser or tenant the notice must state their name and their interest in the works, and in the case of a funder it can simply identify the party concerned. Where collateral warranties are required, the contractor is required to execute the stipulated warranties within 14 days of the equivalent notice from the employer.

2.58   From the point of view of the purchaser, tenant or funder, they will not be aware of the existence of the third party rights unless the employer lets them have a copy of the relevant part of the contract. In some cases the third party may prefer to have a separate collateral warranty direct with the contractor, and it would be sensible of the employer to establish whether this may be a possibility before executing the main contract. After the contract is executed, this could only be arranged with the consent of the contractor.

2.59   With respect to warranties from sub-contractors, the details should also be set out in the Rights Particulars. Clause 7E states that:

> Where the Rights Particulars state that a sub-contractor shall confer third party rights on a Purchaser, Tenant or Funder and/or the Employer or execute and deliver a Collateral Warranty in favour of such person: the Contractor shall comply with the Contract Documents as to the obtaining of such rights or warranties ... .

2.60   It should be noted that this is not, of itself, an absolute requirement for the contractor to obtain the warranties, nor is it even (as with GC/Works1) a requirement to use reasonable endeavours to obtain the warranties – it is simply a requirement to comply with the contract conditions. The contractor is, however, required to include provisions as necessary in sub-contracts in respect of the execution of required warranties (cl 3.9.2.5), and to take 'such steps as are required to obtain each warranty' (cl 7E.1.2). The JCT publishes three standard forms of sub-contract warranty (SCWa/P&T, SCWa/F and SCWa/E) to cover this situation.

3 Obligations of the contractor

3.1 The contractor's paramount obligation is to 'carry out and complete the Works'. This is stated in Article 1 and reinforced in clauses 2.1 and 3.6, the latter being a clear statement that the contractor is held wholly responsible for achieving this, irrespective of whether the contract administrator or clerk of works visits or is present on the site.

3.2 Should Supplemental Provision 1 (Schedule 8) be incorporated, the contractor would, in addition to these primary obligations, be under an express duty of collaboration. The provision states:

> The Parties shall work with each other and with other project team members in a co-operative and collaborative manner, in good faith and in a spirit of trust and respect. To that end, each shall support collaborative behaviour and address behaviour which is not collaborative.

3.3 This places a duty on the contractor to collaborate not only with the employer, but also with other team members, which would include the employer's appointed consultants. Other supplemental provisions introduce further obligations, including to notify the employer promptly of any matter that may give rise to a dispute. Such obligations may affect the interpretation of the nature and extent of the contractor's duties under other clauses.

The works

3.4 The works that the contractor undertakes to carry out will be as briefly described in the first recital of SBC16, and as shown or described in the contract documents. It is therefore important to check that the entry in the first recital clearly identifies the nature and scope of the proposed work, and that descriptions of the works given elsewhere are clear and adequate. Under the ninth recital the employer may stipulate that the works include the design and construction of an identified part or parts of the project, termed 'the Contractor's Designed Portion'. The term 'Works' is consequently defined under clause 1.1 as 'including, where applicable, the CDP Works'.

3.5 Note also that, as defined in clause 1.1, the works will also include any changes subsequently brought about by a contract administrator's instruction, which might also introduce additional drawings or other information. These might not be 'Contract Documents', but they nevertheless have an important status and the contractor is obliged to carry out any additional work which they show.

Contractor's design obligation

3.6 The contractor's design obligation is set out under clause 2.2, which states 'Where the Works include a Contractor's Designed Portion, the contractor shall … in accordance with the Contract Documents, complete the design for the Contractor's Designed Portion'.

3.7   The part or parts to be designed by the contractor are to be identified in the ninth recital, which could if necessary refer to a separate schedule listing the parts. The design requirements will have been set out in the employer's requirements and sent out with the tender documents (tenth recital). The contractor will have submitted a proposal containing a design solution with its tender (the contractor's proposals, eleventh recital) although, depending on the information requested, this may not be fully detailed. Some of the design may therefore remain to be finalised after the contract is entered into.

3.8   Clause 2.13.2 makes it clear that the contractor is not responsible for the contents of the employer's requirements, or for verifying the adequacy of any design contained within them. This clause is included to prevent such an obligation being implied, as it was in the case of *Co-operative Insurance Society* v *Henry Boot*. Although it is not entirely clear, it is unlikely to prevent the implication of a 'duty to warn' regarding any other aspects of the consultant team's design, for example where the design is varied through an instruction (for an example of this see the earlier case of *Plant Construction* v *Clive Adams*).

> *Co-operative Insurance Society* v *Henry Boot Scotland and others* (2002) 84 Con LR 164
>
> The Co-operative Insurance Society (the Society) engaged the contractor Henry Boot on an amended version of JCT80 incorporating the Designed Portion Supplement, where the relevant terms are virtually identical to those of WCD98. During construction, problems arose where soil and water flooded into a basement excavation. An engineer had originally been employed by the Society to prepare a concept design for the structure, and Henry Boot had developed the design and prepared working drawings. The Society brought claims against Henry Boot and the engineers. Henry Boot argued that their liability was limited to the preparation of the working drawings. The judge, however, took the view that completing the design of the contiguous bored pile walls included examining the design at the point that it was taken over, assessing the assumptions on which it was based and forming a view as to whether they were appropriate.

> *Plant Construction* v *Clive Adams Associates and JMH Construction Services* [2000] BLR 137 (CA)
>
> Ford Motor Company engaged Plant on a JCT WCD contract to design and construct two pits for engine mount rigs at Ford's research and engineering centre in Essex. Part of the work included underpinning an existing column, and in the course of the work temporary support was required to the column and the floor above. JMH was sub-contracted to carry out this concrete work. Ford's own engineer gave instructions regarding the temporary supports, which comprised four Acrow props. JMH and Plant's engineers, Clive Adams Associates, felt the props to be inadequate and discussed this on site. The support was installed as instructed and failed, so that a large part of a concrete floor slab collapsed. Plant settled with Ford, and brought a claim against JMH and Clive Adams (who settled). The court found that the duties of the sub-contractor included warning of any aspect of the design that it knew to be unsafe. It reserved its opinion on whether the duty would extend to unsafe aspects it ought to have known about, or design errors that were not unsafe.

3.9   As discussed above, the twelfth recital states 'the Employer has examined the Contractor's Proposals and, subject to the Conditions, is satisfied that they appear to meet the Employer's Requirements', which implies that in so far as the design has been finalised at the time of acceptance of tender, then the employer has accepted the solution. It might be possible to argue that the employer could not be held to have accepted defects in the

Table 3.1 Watchpoints: contractor's design
• The contractor's liability for providing the contractor's designed portion is limited to the use of reasonable skill and care (cl 2.19.1) • The level of professional indemnity insurance must be stated in the contract particulars, otherwise none will be required (cl 6.15) • It is unclear what level of design responsibility the contractor will have for any design not stated to be included in the contractor's designed portion, or whether the contractor is required to insure for this • In cases where the contractor's proposals are found not to comply with the employer's requirements, it is unclear which takes precedence (twelfth recital, cl 2.16) • Integration of the design work remains the responsibility of the contract administrator (cl 2.2.2) • The tender documents should state the exact scope and format of the information to be included in the contractor's proposals • The contractor is obliged to submit further information that it prepares in relation to the contractor's designed portion 'in sufficient time' (Schedule 1:1). It is suggested that the exact information required and dates for submission are set out in the contract documents • Information required to be submitted at practical completion ('as-built' drawings and maintenance information) should be set out in the contract documents (cl 2.40) • The employer may make changes to the employer's requirements (which may result in changes to the contract sum and the completion date), but otherwise there is no power to order changes to the contractor's proposals provided they comply with the employer's requirements (Schedule 1)

design which a reasonable inspection would not have revealed. An example might be the design of a roof truss, where without a detailed double-checking of calculations it would not be possible to ascertain whether the truss would be structurally sound. The contractor would therefore remain responsible for achieving this whatever the contractor's proposals showed. Nevertheless, the recital is problematic from the point of view of the employer and is sometimes deleted. (For a list of watchpoints, see Table 3.1.)

Extent of design liability

3.10   The contractor is only required to design the parts of the project identified in the ninth recital. It is therefore essential that those parts are described clearly and accurately. Leaving this to be agreed or resolved later will undoubtedly lead to problems (see *Walter Lilly & Co. v Giles Mackay & DMW Ltd*). Defining the extent of the contractor's designed portion (CDP) can be quite difficult in practice, especially as it could be several parts or elements, and could also be a system (e.g. services) that is integral to many parts of the building. The contract is clear that the administrator remains responsible not only for any integration, and this could extend to the physical junctions between the CDP and other parts, but also for the combined performance of several systems, or systems with elements. If any of these interfaces (physical or performance) is intended to be the contractor's responsibility to resolve, then the interface would have to be placed firmly within the CDP.

Walter Lilly & Co. v *Giles Mackay & DMW Ltd* [2012] EWHC 649 (TCC)

This was a project for three luxury houses, which began on site with very little finalised design information. The reason for the rushed start was that planning permission had been obtained in 1999 and was subject to a condition that work had to start within five years (i.e. by 15 June 2004).

Apart from the preliminary cost items, all the actual building work was covered by provisional sums. The recitals indicated that the employer's requirements for the CDP were to be 'as notified by the Employer to the Contractor in writing' and the specification stated that 'the following works may be designed by the Contractor', with a list of around 21 possible CDP items. In addition, the specification stated that 'certain Sub-contractors as defined in the Contract will be required to provide design, coordination, fabrication, installation [...] drawings, design calculations, fixing details, specifications and other information as appropriate during the course of the Contract', giving a list of 18 possible types of work, but not indicating any particular firms.

Ultimately, the practical completion certificate was not issued until August 2008, around two years after the original completion date, and even then several key items were omitted from the contract, to be completed later. A major dispute arose over the extent of delays and liability for liquidated damages. Crucial to determining liability was the question of who was responsible for the design of the most defective items, and hence for the consequential delays. Ultimately, the court decided that the contractor was not liable for the design of the doors, nor of any other of the crucial defective elements. The fact that a firm had been involved in developing the design, and the contractor had in some cases entered into sub-contracts where the firm undertook to develop the design, had no bearing on whether the main contractor was liable. If it was intended that Walter Lilly was to take on such responsibility, this should have been made clear in the original contract documents, or through subsequent instructions issued in conformity with the contractual terms.

3.11   Contract administrators sometimes attempt to place a design obligation on the contractor through clauses in the bills or by a reference in other documents, or through the inclusion of a performance specification in a description of the works, other than the contractor's designed portion. It would be unwise to try to assign a design role in this way as the outcome cannot be predicted with certainty. The wording of clause 2.3.3 gives some support to the argument that the contractor would be responsible for providing something 'appropriate to the Works' (see paragraph 3.21 below) and there have been cases where a court has found that ad hoc methods have placed a design obligation on the contractor. However, these did not involve a form which made provision for a 'Contractor's Designed Portion'. In *National Museums and Galleries on Merseyside* v *AEW Architects and Designers Ltd* it was made clear that the contractor's design obligation was limited to the CDP.

National Museums and Galleries on Merseyside v *AEW Architects and Designers Ltd* [2013] EWHC 2403 (TCC)

This project, let on SBC05, was for a new museum, constructed between 2007 and 2011. A key design element was a series of 'half amphitheatre' pre-cast concrete steps and seats at the north and south ends of the museum.

Unfortunately, architects AEW made several errors in co-ordinating the detailed design of the project, including the valley junction between the concrete steps and seats. The steel substructure to these had been redesigned by the engineers (Buro Happold) in August 2007, and AEW failed to appreciate the implications this had for the geometry of the interface between the steps and seats, or to specify the dimensional tolerances between the pre-cast units, or an adequate coverage for the reinforcement to the units, even after they were alerted to the problems in 2008 by a query from the contractor. As a result, it was not possible to use the steps at the time the museum was opened to the public in 2011. The problems resulted in a claim by the museum against the architects, who tried to argue that this detail was the contractor's responsibility. However, the judge would not accept this argument, stating (at para. 82):

> In relation to the gaps, AEW suggests that the design of the steps and seats was part of the works which the Contractor was required to design. This is, simply, wrong. The construction contract identifies those parts of the Works which the Contractor was required to design or have design involvement with as: 'steelwork connections, reinforcement placement & scheduling, general glazing & curtain walling, roof cladding, fixing wind posts, structural glass and glazing'. This is described in the contract as the 'Contractor's Designed Portion' and it is simply in relation to those works that the Contractor has any design responsibility.

Level of design liability

3.12   Standard forms of contract and appointment will often set out specific provisions regarding design liability, but these have to be understood in the legal context in which they operate. A key point is whether any design liability incurred is a 'fitness for purpose' or 'reasonable skill and care' level of liability.

3.13   The Sale of Goods Act 1979 implies terms into all contracts for the sale of goods that the goods sold will be of satisfactory quality. The Consumer Rights Act 2015 stipulates that this requirement cannot be excluded in any contract with a consumer, and under the Unfair Contract Terms Act 1977 it can only be excluded in other contracts in so far as it would be reasonable to do so. If parties have included terms which purport to exclude this liability, the terms will be void. Similarly, if the use to which the goods are to be put is made clear to the seller, the seller must supply goods suitable for that use unless it is clear that the buyer is not relying on the seller's skill and judgement. So if, for example, a DIY enthusiast asks a builder's merchant for paint suitable for use on a bathroom ceiling, the merchant must supply suitable paint, regardless of what is written in the contract of sale. If, however, the buyer specifies the exact type of paint, the seller would no longer be liable as the buyer is not relying on the seller's advice.

3.14   Contracts for construction work are usually for 'work and materials' (as opposed to supply-only or install-only) and as such fall under the Supply of Goods and Services Act 1982. This implies similar terms to those described above in relation to any goods supplied under such a contract. Therefore, a contractor would normally be liable for providing materials fit for their intended purposes. If, however, an employer or consultant specifies particular materials, the contractor would be relieved of this liability.

3.15   The obligation to supply goods or materials fit for their intended purpose would extend to a product or structure which a contractor had agreed to design and construct (*Viking Grain Storage Ltd* v *T H White*). In all cases the liability of the contractor will be strict; in other words, the contractor will be liable if the goods, element or structure is not fit for its intended use, irrespective of whether the contractor has exercised a reasonable level of skill and care in carrying out the design. This is a more onerous level of liability than that assumed by someone undertaking design services only, where they would normally be required to demonstrate that they had exercised the skill and care of a competent member of their profession. To put it the other way around, if an employer can prove that a building designed and constructed by a contractor is defective, then this will normally be sufficient to prove that there has been a breach of contract, whereas in the case of a design professional, the employer would also have to prove that the professional had been negligent.

> *Viking Grain Storage Ltd* v *T H White Installations Ltd* (1985) 33 BLR 103
>
> Viking Grain entered into a contract with White to design and erect a grain drying and storage installation to handle 10,000 tonnes of grain. After it was complete, Viking commenced proceedings against the contractor claiming that, because of defects, the grain store was unfit for its intended use. The contractor, in its defence, claimed that there was no implied warranty in the contract that the finished product would be fit for purpose, and that the contractor's obligation was limited to the use of reasonable skill and care in carrying out the design. The judge decided that Viking had been relying on the contractor and, because of this reliance, there was an implied warranty that, not only the materials supplied, but also the whole installation should be fit for the required purpose. There could be no differentiation between reliance placed on the quality of the materials and on the design.

3.16   Under clause 2.19.1 of SBC16, the contractor's liability for the contractor's designed portion is equivalent to that of 'an architect or other appropriate professional designer who holds himself out as competent to take on work for such design'. In effect, this means that in order to prove a breach the employer would need to prove that the contractor had been negligent. If, for example, the contractor is required to design a heating system to heat the rooms to a certain temperature, and when installed the system fails to do so, this fact alone would not be enough to prove that there had been a breach of contract. The employer would need to prove that the contractor had failed to use the skill and care expected of a professional person.

3.17   It should be noted that where the contractor is carrying out work in connection with a dwelling, including design work, this would be subject to the Defective Premises Act 1972. This obligation is acknowledged in clause 2.19.2. The Act states that 'A person taking on work for or in connection with the provision of a dwelling … owes a duty … to see that the work which he takes on is done in a workmanlike or, as the case may be, professional manner, with proper materials and so that as regards that work the dwelling will be fit for habitation when completed' (section 1(1)). In case law the duty has not generally been taken to be a strict or absolute warranty of fitness (*Alexander* v *Mercouris*), although other authorities suggest that it is a strict duty. In addition, it should be noted that although the contractor's liability is limited to the amount stated in the contract particulars, the limitation does not apply to work in connection with a dwelling (cl 2.19.3).

> *Alexander* v *Mercouris* [1979] 1 WLR 1270
>
> This case considered when the duty arose, not its scope, but some observations are helpful, for example Lord Justice Buckley stated (at page 1274):
>
>> It seems to me clear upon the language of Section 1(1) that the duty is intended to arise when a person takes on the work. The word 'owes' is used in the present tense and the duty is not to ensure that the work has been done in a proper and workmanlike manner with proper materials so that the dwelling is fit for habitation when completed, but to see that the work is done in a proper and workmanlike manner with proper materials so that the work will be fit for habitation when completed. The duty is one to be performed during the carrying on of the work. The reference to the dwelling being fit for habitation indicates the intended consequence of the proper performance of the duty and provides a measure of the standard of the requisite work and materials. It is not, I think, part of the duty itself.

Materials, goods and workmanship

3.18   Work must be carried out in a proper and workmanlike manner and in accordance with the construction phase plan (cl 2.1). Any failure in these respects is regarded as serious, and the contract administrator has the power to intervene by issuing an instruction if necessary (cl 3.19).

3.19   Clause 2.3.1 states that all materials and goods shall be of the standard described in the contract bills (or in the specification in the Without Quantities version). The obligation is qualified by the phrase 'so far as procurable'. It would be an implied duty that the contractor should notify the contract administrator before substituting any materials or goods, even where those specified are unobtainable. The substitution would result in a variation that should be covered by a contract administrator's instruction. With respect to the contractor's designed portion, the standard of materials and goods should be that in the employer's requirements or, if none is given, then that shown in the contractor's proposals or other contractor's design documents. Unlike DB16, there are no clauses in SBC16 referring to an obligation to provide samples (see DB16 cl 2.2.3). If samples are required, then additional provisions would need to be included in the contract.

3.20   Standards of workmanship should be as specified in the contract bills (or specification, cl 2.3.2). In respect of the contractor's designed portion, the standard of workmanship is that set out in the employer's requirements or, if none is set out, is that set out in the contractor's proposals.

3.21   Clause 2.3.3 states:

> Insofar as the quality of materials or goods or standards of workmanship are stated to be a matter for the Architect/Contract Administrator's approval, such quality and standards shall be to his reasonable satisfaction. To the extent that the quality of materials and goods or standards of workmanship are neither described… nor stated to be a matter for such approval or satisfaction, they shall in the case of the Contractor's Designed Portion be of a standard appropriate to it and shall in any other case be of a standard appropriate to the Works.

3.22   This reflects the duty that would normally be implied by law. In other words, where the description of the standard required for any goods, materials and workmanship is (deliberately or inadvertently) incomplete, the contractor is required to provide something 'fit for purpose' (see paragraph 3.13). This appears to be a strict obligation (see above), rather than an obligation to use reasonable skill and care.

3.23   The phrase 'Insofar as the quality… are stated to be a matter for the Architect/Contract Administrator's approval, such quality and standards shall be to his reasonable satisfaction' (cl 2.3.3) does not authorise the contract administrator to alter the standard specified at will, but means that where a correct construction of the contract documents leaves a matter regarding quality to the approval of the contract administrator, the contractor only fulfils its obligations if the contract administrator is satisfied. Any expression of dissatisfaction by the contract administrator must be made within a reasonable period of the carrying out of the unsatisfactory work, and the contract administrator must state reasons for the dissatisfaction (cl 3.20). It should be noted that the final certificate is conclusive evidence that where the contract documents have expressly stated that the quality is to be to the approval of the contract administrator, then the contract administrator is so satisfied (cl 1.9.1). This would have the effect of preventing the employer bringing a claim regarding those items of work. The contract administrator should generally avoid using phrases such as 'to approval' or 'to the contract administrator's satisfaction' in the contract documents.

3.24   If the phrase 'or otherwise approved' is used in a contract bill or specification this does not mean that the contract administrator must be prepared to consider alternatives put forward by the contractor, nor that the contract administrator must give any reasons for rejecting alternatives (*Leedsford* v *City of Bradford*). It merely gives the contract administrator the right to do so. A substitution would always constitute a variation whether or not this phrase is present in the specification.

> *Leedsford Ltd* v *The Lord Mayor, Alderman and Citizens of the City of Bradford* (1956) 24 BLR 45 (CA)
>
> In a contract for the provision of a new infant school the contract bills stated 'Artificial Stone ... The following to be obtained from the Empire Stone Company Limited, 326 Deansgate, or other approved firm'. During the course of the contract the contractor obtained quotes from other companies and sent them to the architect for approval. The architect, however, insisted that Empire Stone was used and, as Empire Stone was considerably more expensive, the contractor brought a claim for damages for breach of contract. The court dismissed the claim stating: 'The builder agrees to supply artificial stone. The stone has to be Empire Stone unless the parties agree some other stone, and no other stone can be substituted except by mutual agreement. The builder fulfils his contract if he provides Empire Stone, whether the Bradford Corporation want it or not; and the Corporation Architect can say that he will approve of no other stone except the Empire Stone' (Hodson LJ at page 58).

Obligations in respect of quality of sub-contracted work

3.25   With increasing specialisation in the construction industry it is almost universal practice for much of the work on building projects to be sub-contracted to a large number of other firms. This arrangement benefits the employer by enabling it to take advantage of a wider range of specialisms than would normally be available within one contracting organisation. The employer will wish, nevertheless, to be able to hold the contractor responsible to a degree for any non-performance of the sub-contractors.

3.26   There are three methods of sub-contracting allowed for under SBC16:

- sub-letting to a domestic sub-contractor selected by the main contractor, but with the written consent of the contract administrator (cl 3.7);

- sub-letting to a domestic sub-contractor selected from a list of at least three names under the procedure set out in clause 3.8;

- sub-letting to a 'named specialist' sub-contractor under the procedure set out in Supplemental Provision 9.

3.27   All of these methods are discussed in detail in Chapter 5. With respect to quality, in all cases the contract makes it clear that the contractor remains entirely responsible for the performance of any sub-contractor (cl 3.7.1). This responsibility would extend to any contractor's designed portion work that was sub-contracted (although note that the named specialist provisions are not intended to be used for contractor's designed portion work). As discussed at paragraph 3.11, if any ad hoc method is adopted, such as attempting to 'nominate' a sub-contractor in the contract bills or specification, then responsibility for quality and fitness will depend on the particular facts of each case. It should be noted, however, that attempts to hold the contractor liable based on early editions of the standard form were not always successful (see for example *Gloucestershire County Council* v *Richardson*).

> ***Gloucestershire County Council* v *Richardson* [1969] 1 AC 480 (CA)**
>
> Under a building contract on JCT37 (1957 edition), the contractor was to supply concrete columns, which were to be ordered from suppliers nominated by the employer. A prime cost sum had been stated in the bills of quantities, which named a particular supplier. On accepting the tender, the architect nominated a different supplier, who it transpired would only contract on particular terms, which excluded all liability for defects or their damaging consequences save only an obligation to replace. The columns had defects which were undetectable when they were supplied, but which appeared after some columns had been incorporated in the building under construction. Once the defects were discovered the contractor was told to stop work on the columns, which eventually led to the contractor terminating the contract.
>
> In an action by the employer against the contractor, the employer argued that the contractor had repudiated the building contract, and that the defects in the concrete columns were a breach of its implied warranty of fitness. The court found that the contractor was not liable for the defects, commenting that it felt that to find otherwise would be unfair. Russell LJ pointed out that 'if the employers wished to impose liability on the contractor for defects in materials supplied by nominated suppliers they could do so in plain terms, such as are contained in clause 31 of the General Conditions of Government Contracts (October 1959)'.

Compliance with statute

3.28   The contractor is under a statutory duty to comply with all legislation that is relevant to the carrying out of the works, for example in respect of goods and services, building and construction regulations, and health and safety. The duty is absolute and there is no possibility of contracting out of any of the resulting obligations.

3.29   SBC16 imposes a contractual duty in addition to the statutory duty, which gives additional protection to the employer, in that failure to comply with statute becomes a breach of contract. Under clause 2.1 the contractor is required to carry out and complete the works 'in compliance with … the Construction Phase Plan and other Statutory Requirements'. The contractor is required to pay all fees and charges, but unless these have already been included in the contract sum or are covered by a provisional sum, the amounts will be added to the contract figure (cl 2.21).

3.30   If the contractor finds any divergence between what the contract requires and statutory requirements, then the contract administrator must be given immediate written notice and, where the divergence is between the statutory requirements and the employer's requirements, the contractor's proposals or the contractor's design documents, must set out its proposals for dealing with the divergence (cl 2.17.1). Provided the contractor complies with this requirement, it is not liable for any non-compliant work, other than the CDP works (cl 2.17.3). Once either the contractor or the contract administrator discovers a divergence, the contract administrator must issue an instruction within seven days to rectify the situation.

This is treated as a variation (cl 2.17.2), and may therefore give rise to an extension of time and reimbursement of direct loss and/or expense (cl 2.29.1 and 4.22.1).

3.31   The contractor may need to take immediate action in an emergency, but only in so far as is reasonably necessary to comply with statutory requirements (cl 2.18). Unless it is an emergency, any alteration made by a contractor, for example at the request of a district surveyor, could be a breach of contract, unless the contract administrator decides to sanction the variation under clause 3.14.4.

3.32   Under clause 3.23 each party undertakes to the other to comply with all their obligations under the Construction (Design and Management) (CDM) Regulations 2015. The principal designer and principal contractor are named under Articles 5 and 6 (the contract assumes that the contractor will act as principal contractor, unless another firm is named). Clause 3.23.1 places a contractual obligation on the employer to ensure that the principal designer carries out his or her duties under the CDM Regulations. There are equivalent provisions where the contractor is not the principal contractor. This is a wider obligation than the 'reasonable satisfaction with competence' obligation imposed by the Regulations. Breach of this clause gives the contractor the right to terminate its employment under the contract under clause 8.9.1.4. It is more likely, in practice, that the contractor will claim for an extension of time or direct loss and/or expense for breach of clause 3.23, as this is a 'Relevant Event' under clause 2.29.7 and a 'Relevant Matter' under clause 4.24.5. An example might be where a principal designer delays in commenting on a contractor's proposed amendment to the construction phase plan, and progress is thereby delayed.

3.33   Clause 3.23.2 places a duty on the contractor to comply with regulations 8 and 15 and, if acting as the principal contractor, to comply with regulations 12 to 14. Breach of this duty is grounds for termination under clause 8.4.1.5. The warning notice still has to be given, and JCT Practice Note 27 suggested that the provision should only be used for situations where the Health and Safety Executive is likely to close the site. Any breach is covered, however, provided termination is not unreasonable or vexatious, and the employer should consider any breach that might lead to action being taken against it to be a serious one. If work needs to be postponed or other instructions given due to a breach by the contractor, then there should be no entitlement to an extension of time or a claim for direct loss and/or expense.

3.34   The contractor should take the cost of compliance with the CDM Regulations (e.g. the cost of developing the construction phase plan) into account at the tender stage. No claims may be made for compliance with the Regulations (e.g. adjusting the construction phase plan to suit the contractor's or sub-contractor's working methods) and no extension of time will be given (cl 3.23.3). If alterations are needed as a result of a variation instruction, then the costs are included in valuing the variations and the alterations may be taken into account in assessing an application for an extension of time.

3.35   The CDM Regulations are not the only statutory health and safety obligations which may apply to the project, and the contractor must comply with all applicable health and safety laws. Furthermore, Supplemental Provision 2 (Schedule 8) states that 'the Parties will endeavour to establish and maintain a culture and working environment in which health and safety is of paramount concern', which suggests that a 'best practice' rather than a minimum compliance approach is required. The provision sets out several specific requirements; for example, the contractor undertakes to comply with all approved codes of practice, to ensure that personnel receive induction training and have access to advice and to ensure that there is full and proper health and safety consultation with all such personnel in accordance with the Health and Safety (Consultation with Employees) Regulations 1996.

Other obligations

3.36   In addition to the major obligations outlined above, the contractor also has other obligations arising out of the contract. The most significant of these are in relation to progress and programming, discussed in the following chapter, and in regard to insurance matters, discussed in Chapter 8. The contractor's obligations and powers are summarised in Tables 3.2 and 3.3.

Table 3.2	Key duties of the contractor
Clause	**Duties of the contractor**
1.7	Communicate notices, etc. by agreed means
1.11	Ensure consents, approvals, etc. not unreasonably delayed or withheld
2.1	Carry out and complete the works
2.2.1	Complete the design of the CDP
2.2.2	Comply with the contract administrator's directions regarding integration of the CDP
2.2.3	Comply with regulations 8–10 of the CDM Regulations
2.3.1	Provide materials and goods of standards described
2.3.2	Provide workmanship and goods of standards described
2.3.3	Provide materials, goods and workmanship to satisfaction of contract administrator
2.3.4	Provide the contract administrator with vouchers
2.3.5	Take all reasonable steps to encourage contractor's persons to be CSCS registered cardholders
2.4	Retain possession of the site
2.6	Notify the insurers of early occupation; notify the employer of any additional insurance premium
2.7.1	Permit execution of work by employer's persons
2.8.3	Keep copy of documents on site
2.9.1.2	Provide copies of master programme
2.9.2	Provide copies of amended master programme
2.9.4	Provide copies of contractor's design documents in accordance with design submission procedure
2.15	Notify the contract administrator of discrepancies
2.16.1	Notify the contract administrator of discrepancies in relation to CDP documents
2.16.2	Inform the contract administrator of amendment to deal with discrepancies
2.17.1	Notify the contract administrator of any discrepancy between documents and statutory requirements

Table 3.2 Key duties of the contractor – Continued

Clause	Duties of the contractor
2.18.1	Execute work in relation to emergency compliance with statutory requirements
2.18.2	Inform the contract administrator of cl 2.18.1 work
2.21	Pay and indemnify the employer against liability for statutory fees
2.22	Indemnify the employer against liability for royalties and patent rights
2.23.2	Notify the contract administrator of potential infringement of patent rights
2.27.1	Notify the contract administrator of delays and their causes
2.27.2	Notify the contract administrator of further particulars of delays
2.27.3	Notify the contract administrator of any material change in the estimated delay
2.38	Make good defects notified by the contract administrator
2.40	Supply contractor's design documents and related information (as-built drawings)
3.1	Secure a right of access for the contract administrator to sub-contractors' workshops
3.2	Appoint a full-time site manager, to be on site at all times (or a competent deputy)
3.9	Engage sub-contractors using the relevant version of the JCT sub-contract, or on terms that incorporate the provisions set out in clause 3.9
3.10	Comply with instructions issued by the contract administrator
3.10.3	Notify the contract administrator of the adverse effect of an instruction
3.12.1	Confirm oral instruction
3.22.1.1	Use best endeavours not to disturb any object of antiquity
3.22.1.2	Take all steps necessary to preserve the object
3.22.1.3	Inform the contract administrator of discovery of object
3.23	Comply with CDM Regulations
3.23.2	Comply with CDM regulations 8 and 15 and, if principal contractor, 12–14
4.7	Provide a bond in relation to any advance payment
4.11.2	Pay employer amount stated in certificate
4.11.5	Issue pay less notice to employer if intending to pay less than the certified sum
4.11.6	Pay interest to the employer on unpaid amounts
4.16.2	Provide proof regarding ownership, etc. of off-site materials
4.16.4 and 5	Provide a bond in relation to off-site materials
4.18.2	Provide a bond in lieu of retention
4.18.4	Arrange for increase in amount stated in bond
4.21.1	Notify the contract administrator as soon as the likely effect of a relevant matter is known
4.21.2	Provide supporting information in relation to loss and/or expense
4.21.3	Update the contract administrator at monthly intervals in relation to loss and/or expense
4.25.1	Provide the contract administrator with documents for final adjustment of the contract sum

Table 3.2 Key duties of the contractor – Continued

Clause	Duties of the contractor
5.3.1	Provide a Schedule 2 quotation
6.1	Indemnify the employer against losses, etc. due to personal injury or death of any person
6.2	Indemnify the employer against losses, etc. due to damage to property other than the works
6.4.1	Take out and maintain insurance in respect of cl 6.1 and cl 6.2 liability
6.5.1	If required under the contract particulars, take out and maintain insurance in respect of damage to property caused (without negligence) by carrying out of works
6.5.2	Deposit the cl 6.5.1 policy with the contract administrator
6.7.2	Maintain works insurance up to practical completion
6.9.1	Ensure joint names policy provides for recognition of, or waives any right of subrogation against, any sub-contractor
6.10.1	Take out terrorism cover
6.11.1	Notify the employer if notified that terrorism cover has ceased
6.12.1	Provide evidence of insurance to the employer
6.13.1	Notify the contract administrator and the employer of loss or damage to the works or existing structure
6.13.3	Authorise insurers to pay all monies to employer
6.13.4	Restore all damaged work, etc.
6.15	Take out and maintain professional indemnity insurance, and provide evidence that insurance has been effected
6.16	Give notice to employer if cl 6.15 insurance ceases to be available
6.18	Comply with the Joint Fire Code, and ensure all contractor's persons comply
6.19	Ensure remedial measures are carried out
7.3	Provide the employer with a performance bond or guarantee
7E	Comply with requirements set out in contract particulars relating to obtaining sub-contractor warranties for purchasers, tenants/funders or employer
8.5.2	Give the employer notice if it makes any proposal, etc. in relation to insolvency
8.7.2.1	Remove temporary buildings, etc. from the site
8.7.2.2	Provide employer with copies of all design documents
8.7.2.3	Assign to the employer the benefit of any agreements to supply materials, etc. or execute work
8.7.5	Pay the employer any balance due
8.12.2.1	Remove temporary buildings, etc. from the site
8.12.2.2	Provide employer with copies of all design documents
8.12.3	Prepare an account or provide employer with documents to prepare an account
9.1	Give serious consideration to a request from the employer to refer a dispute to mediation

Table 3.2 Key duties of the contractor – Continued

Clause	Duties of the contractor
Schedule 1	
1	Prepare and submit two copies of each design document to the contract administrator
5.1	Carry out CDP works
5.2	Incorporate comments and carry out CDP works
Schedule 2	
1.2	Submit Schedule 2 quotation
Schedule 3	
A.1	Effect and maintain a joint names policy for 'All Risks Insurance'
Schedule 4	Endeavour to agree the amount and method of opening up and testing
Schedule 7	
A.3.1	Incorporate specified provisions in any sub-contract
A.4.1	Notify the contract administrator of occurrence of events within reasonable period of time
A.7	Provide such evidence and computations as the contract administrator or quantity surveyor may reasonably require
Schedule 8	
1	Work in a collaborative manner
2.1	Endeavour to create a working environment where health and safety is of paramount concern
2.2	Comply with HSE codes, ensure personnel and supply chain have health and safety training and access to advice, ensure there is proper health and safety consultation
3.2	Provide details of proposed changes
3.3	Negotiate with a view to agreeing consequences of a proposed change
4.2	Provide details of environmental impact
5.2	Provide information necessary to monitor contractor's performance against indicators
6	Promptly notify the employer of matters likely to give rise to a dispute, and meet as soon as possible for good faith negotiations
8	Where the employer is a local or public authority, include specific provisions in any sub-contract and take steps to terminate a sub-contractor's employment if required to do so by the employer
9.2	Enter into a sub-contract with a named specialist
9.5	Notify the contract administrator if intending to terminate the employment of a named specialist

Note: CDP = contractor's designed portion

Table 3.3 Key powers of the contractor

Clause	Powers of the contractor
2.6.1	Consent to early use by employer
2.7.2	Consent to work being carried out by others
2.33	Consent to partial possession by employer
3.7	Sub-contract the works, with the contract administrator's consent
3.8.1	Add persons to the list of sub-contractors
3.10.1	Make a reasonable objection to an instruction
3.13	Request the contract administrator to specify provision empowering an instruction
4.10.1	Submit a payment application to the quantity surveyor
4.10.2.2	Give a payment notice to the quantity surveyor
4.11.5	Issue a pay less notice
4.13	Suspend performance of its obligations
5.3.1	Disagree with the application of the Schedule 2 procedure
6.12.2	Take out insurance in event of employer's apparent failure to do so
6.14	Terminate its employment
8.9.1	Give notice to employer specifying defaults
8.9.2	Give notice to employer specifying suspension events
8.9.3	Terminate its employment because of continuation of specified default or suspension event
8.9.4	Terminate its employment because of repeat of specified default or suspension event
8.10.1	Terminate its employment because of employer insolvency
8.11.1	Terminate its employment because of suspension of the works
9.4.1	Give notice of arbitration
9.4.3	Give further notice of arbitration
9.7	Apply to the courts to determine a question of law
Schedule 1	
7	Notify the contract administrator that it disagrees with comment on design document, with reasons
Schedule 2	
1.1	Notify the contract administrator that information provided is insufficient to prepare quotation

4 Programme

4.1 The date of possession or commencement and the date for completion of the works are key dates in any building contract and SBC16 requires a 'Date of Possession' and a 'Date for Completion' to be inserted into the contract particulars. SBC16 also offers the facility for the work to be carried out in phases. If phased completion is required, then the work must be split into clearly identified sections, and a separate date of possession and date for completion entered for each section. The contractor is required to take possession on the date of possession and complete by the date for completion, and failure to do so may give rise to liability for liquidated damages.

4.2 It is essential that actual dates are given at the time of tendering, and not vague indications such as 'to be agreed' or 'eight weeks after approval by ... '. It is important that the contractor knows the time of year that it will be carrying out the works, as the start date and duration will affect the tender figure. In the event that work is started without proper agreement on dates, the contract will be subject to the Supply of Goods and Services Act 1982 or the Consumer Rights Act 2015, which state that completion is to be within a reasonable time. If the contractor fails to complete within a reasonable time, the employer will be unable to claim liquidated damages and will have to prove any damages it wishes to recover.

4.3 The contract allows for the employer to defer possession under an optional clause, and for the contract administrator to adjust the dates for completion in certain circumstances. It also allows for the parties to agree an adjustment as a consequence of a variation to the works (a 'Pre-agreed Adjustment', see paragraph 4.11). For a summary of the key milestones in SBC16 see Figure 4.1.

Possession by the contractor

4.4 Possession of the site is a fundamental term of the contract. Failure to give the contractor possession is a serious breach by the employer, which may amount to repudiation, and therefore give the contractor the right to treat the contract as at an end, or entitle it to terminate its employment (cl 8.9.2). Giving possession of only part of the site could amount to a breach unless this intention has been made clear in the contract documents (*Whittal Builders* v *Chester-le-Street DC*).

> **Whittal Builders Co. Ltd v Chester-le-Street District Council** (1987) 40 BLR 82
>
> Whittal Builders contracted with the Council on JCT63 to carry out modernisation work to 90 dwellings. The contract documents did not mention the possibility of phasing but the Council gave the contractor possession of the houses in a piecemeal manner. Even though work of this nature was frequently phased, the judge nevertheless found that the employer was in breach of contract for not giving the contractor possession of all 90 dwellings at the start of the contract, and the contractor was entitled to damages.

Figure 4.1 The SBC16 timeline

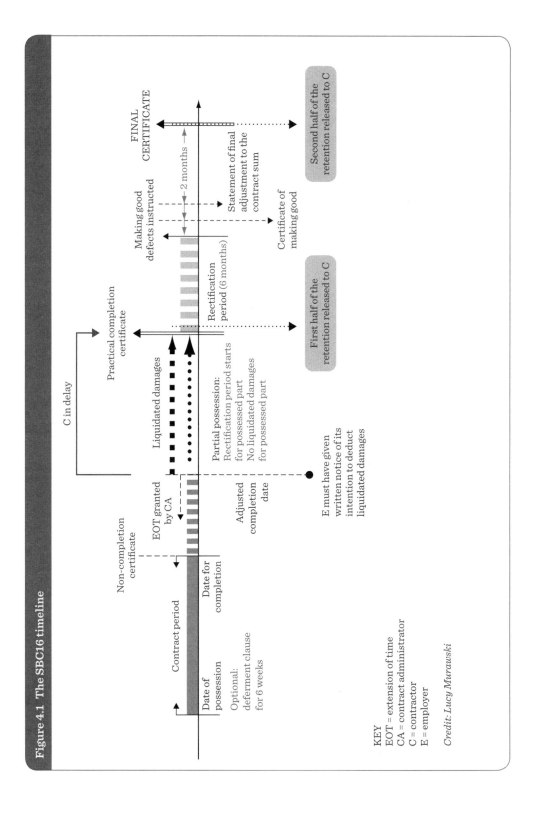

KEY
EOT = extension of time
CA = contract administrator
C = contractor
E = employer

Credit: Lucy Murawski

4.5   SBC16 requires that the contractor is given possession of the site on the date of possession (cl 2.4). Under clause 1.1, a section is defined as a subdivision of the works, not the site, but clause 2.4 clarifies the position, stating 'in the case of a Section, possession of the relevant part of the site shall be given'. In some cases, where the section is all pervasive (e.g. where the mechanical and electrical services have been defined as a section), access to the entire site may be required. Any proposed subdivision of the site in relation to the sections must be made completely clear in the contract documents, together with proposed arrangements for access to each subdivision.

4.6   The contract allows for the employer to defer the giving of possession of the site or of any relevant part under an optional clause (cl 2.5), for a period not exceeding six weeks. In practice this is a very useful provision, as it allows the employer to make small adjustments to the commencement date without having to renegotiate the contract with the contractor. The tender documents must have stated that the clause is to apply, and the relevant section of the contract particulars must be completed. If a period less than six weeks is to be allowed, this must be inserted in the contract particulars. Where the works are split into sections, it is possible to set different periods of deferment for each section, again up to a maximum of six weeks.

4.7   If it becomes necessary to defer possession, the employer must notify the contractor in writing. Although the contract does not require it, it would be wise to do this as far in advance of the planned commencement date as possible. The contractor will be entitled to an extension of time (cl 2.29.3) and loss and/or expense (cl 4.20), and early notification should keep the losses to a minimum. The clause must be operated strictly according to the conditions, and any delay beyond the periods stated in the contract particulars would be a breach of contract.

4.8   The parties are, of course, always free to renegotiate the terms of any contract. Therefore, if there is a delay in giving possession which is longer than the period stated in the contract particulars, or where the contract particulars have stated that clause 2.5 does not apply, the parties would have to agree a new date of possession, usually with a financial compensation to the contractor. Any further delay beyond the agreed date would, of course, be a breach.

4.9   Degree of possession is such that there must be no interference that prevents the contractor from working in whatever way or sequence it chooses. With most jobs this means that the contractor must be given clear possession of the whole site until practical completion. Clause 2.4 supports this by stating that 'the Employer shall not be entitled to take possession of any part or parts of the Works' until the date of issue of the practical completion certificate. Where clear possession is not intended then the tender documents should set out in detail the restrictions and the contract must be amended accordingly. Access to the site and to surrounding areas in the control of the employer should also be made clear, in case this leads to disputes (see *The Queen in Rights of Canada* v *Walter Cabbott Construction Ltd*). Should the employer wish to use any part of the works for any purpose during the time that the contractor has possession, this should also be made clear in the tender documents, otherwise it can only be with the agreement of the contractor (cl 2.6.1). Similarly, should the employer wish the contractor to allow access for others to carry out work, this should also be made clear (cl 2.7.1).

> *The Queen in Rights of Canada* v *Walter Cabbott Construction Ltd* (1975) 21 BLR 42
>
> This Canadian case (Federal Court of Appeal) concerned work to construct a hatchery on a site (contract 1) where several other projects relating to ponds were also planned (contracts 3 and 4). The work to the ponds could not be undertaken without occupying part of the hatchery site. Work to the ponds was started in advance of contract 1, causing access problems to the contractor when contract 1 began. The court confirmed (at page 52) the trial judge's view that 'the "site for the work" must, in the case of a completely new structure comprise not only the ground actually to be occupied by the completed structure but so much area around it as is within the control of the owner and is reasonably necessary for carrying out the work efficiently'.

Progress

4.10   It would normally be implied into a construction contract that a contractor will proceed 'regularly and diligently' with the work, and this is an express term in SBC16 (cl 2.4). The contractor is free to organise its own working methods and sequence of operations, with the qualification that it must comply with statutory requirements and the construction phase plan (cl 2.1). This freedom has been held to be the case even where the contractor's chosen sequencing may cause extra cost to the employer with the operation of fluctuation provisions (*GLC* v *Cleveland Bridge and Engineering*).

> *Greater London Council* v *Cleveland Bridge and Engineering Co.* (1986) 34 BLR 50 (CA)
>
> The Greater London Council (GLC) employed Cleveland Bridge to fabricate and install gates and gate arms for the Thames Barrier. The specially drafted contract provided dates by which Cleveland Bridge had to complete certain parts of the works. Clause 51 was a fluctuations provision which allowed for adjustments to be made to the contract sum if, for example, the rates of wages or prices of materials rose or fell during the course of the contract. The clause also contained the phrase 'provided that no account shall be taken of any amount by which any cost incurred by the Contractor has been increased by the default or negligence of the Contractor'. The contract was lengthy, and Cleveland Bridge left a part of the work to be carried out at the very end of the period, but delivered the gates on time. The result was that the GLC had to pay a large amount of fluctuations in respect of the work done at the last minute. The GLC argued that the contractor had failed to proceed regularly and diligently, and therefore was in default. The court held that even if the slowness of the contractor's progress might at certain points have given the employer the right to terminate the contract under the termination provisions, this would not, in itself, be a breach of contract as referred to in clause 51. The contractor could organise the work in any way it wished, provided it completed on time: it was therefore owed the full amount of the fluctuations. (Note that JCT forms contain a freezing provision which prevents fluctuations from operating when the contractor is in culpable delay.)

4.11   SBC16 requires the contractor to produce a 'master programme' (cl 2.9.1.2). The programme is not a contract document, and the clause expressly states that it does not impose any additional obligations on either party (e.g. it would not oblige the contractor to carry out the work in any particular sequence). The programme may nevertheless be useful to the contract administrator, particularly in regard to monitoring progress and assessing extensions of time. Clause 2.9.1 states that it should be provided 'as soon as possible' after the execution of the contract, if not previously provided. The contractor is also required to issue 'an amendment or revision of the master programme' each time a new completion date is fixed by the contract administrator under clause 2.28.1, or where there is a confirmed acceptance of a Schedule 2 quotation (termed a 'Pre-agreed Adjustment').

4.12   The provisions relating to the programme are brief in comparison with those in many other standard forms, and the employer may wish to consider whether any additional requirements should be included in the tender documents. As in many projects, it is important to have a programme at an early stage and, as the form has no sanction for its non-provision, it may be wise to insist on the programme at tender stage or before executing the contract. Alternatively, if a more specific time limit were required, for example not less than two weeks before the date of possession, this would require an amendment to clause 2.9.1.2.

4.13   No particular format is required by SBC16 for the programme, but again requirements could be set out in the tender documents; for example, that it must show a critical path and/or allocation of resources. If the information release schedule is not used it would be open to the contractor to set out dates when key information will be required, although these would not be contractually binding. It is common practice to ask for a detailed programme indicating activities, a critical path and resources, and possibly key dates when information regarding the developing design will be supplied to the employer. SBC16 does not require the contractor to publish a revised programme when the contractor is in delay, so if regular updates are required a revision to clause 2.9.1.2 should be considered.

4.14   The contract administrator should be most careful never to 'approve' a programme in such a way that it becomes a contract document against which the contract administrator's own performance in providing information will be judged. A programme which shows a large float period between the contractor's estimated completion date and the date entered in the contract particulars should be queried. For example, if the contractor has tendered to carry out the work in 30 months but produces a programme showing estimated completion in 24 months, then the supply of information by the contract administrator might be judged on the 24-month period if the contract administrator appears to have agreed to this. Also, the early completion date may create difficulties for the employer. It is clear, however, that just because a contractor's programme shows an intention to complete early, there is no implied duty on the employer to enable the contractor to achieve this early completion (*Glenlion Construction* v *The Guinness Trust*).

> *Glenlion Construction Ltd* v *The Guinness Trust* (1987) 39 BLR 89
>
> The Trust employed Glenlion Construction to carry out works in relation to a residential development at Bromley, Kent. The contract was on JCT63, which required the contractor to complete 'on or before' the date for completion, and to provide a programme. Disputes arose which went to arbitration and several questions of law regarding the contractor's programme were subsequently raised in court. The contractor later claimed loss and expense on the ground that it was prevented from economic working and achieving the early completion date shown on its programme only by failure of the architect to provide necessary information and instructions by the dates shown. The court decided that Glenlion was entitled to complete before the date for completion, whether or not it was contractually bound to produce a programme and whether or not it did in fact produce one. Glenlion was therefore entitled to carry out the works in a way which would achieve an earlier completion date. However, there was no implied obligation that the employer (or the architect) should perform its obligations so as to enable the contractor to complete by any earlier completion date shown on the programme.

Completion

4.15   Building contracts normally stipulate a completion date for the works. The importance of this completion date is that it provides a fixed point from which damages may be payable in the event of non-completion. Generally in construction contracts the damages are 'liquidated', and typically are fixed at a rate per week of overrun.

4.16   The contractor is obliged to complete the works by the completion date, and in general accepts the risk of all events that might prevent completion by this date. The contractor is relieved of this obligation if the employer causes delays or in some way prevents completion. In addition, most contracts contain provisions allowing for the adjustment of the completion date in the event of certain delays caused by the employer, or neutral delaying events. The contract dates can, of course, always be adjusted by agreement.

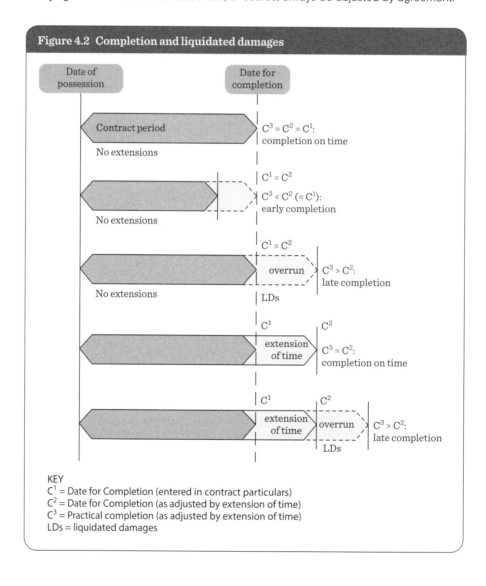

Figure 4.2 Completion and liquidated damages

KEY
C^1 = Date for Completion (entered in contract particulars)
C^2 = Date for Completion (as adjusted by extension of time)
C^3 = Practical completion (as adjusted by extension of time)
LDs = liquidated damages

4.17   In contracts it is sometimes essential that completion is achieved by a particular date and failure would mean that the result is worthless. This is sometimes referred to as 'time is of the essence'. Breach of such a term would be considered a fundamental breach, and would give the employer the right to terminate performance of the contract and treat all its own obligations as at an end. The expression 'time is of the essence' is seldom, if ever, applicable to building contracts such as SBC16, as the inclusion of extensions of time and liquidated damages provisions implies that the parties intended otherwise.

4.18   SBC16 uses the following terms:

- 'Date for Completion' – the date inserted in the contract particulars, which is the date agreed at the time of entering into the contract. Where the works are divided into sections, a separate date will be stated for each section;

- 'Completion Date' – the date for completion of the works or a section, or any later date consequent upon an extension of time or a pre-agreed adjustment;

- 'practical completion' – the date at which, in the opinion of the contract administrator, the works or a section are complete.

4.19   The form provides for the granting of extensions of time, which result in the fixing of a new completion date (for the works, or any section). However, the form makes no provision for the contract administrator to reduce the contract period to less than that stated in the contract particulars, even when work is omitted (cl 2.28.6.3). Under the Schedule 2 quotation procedure (see below) the employer may agree an extension or reduction to the contract period with the contractor, including reducing it to less than that stated in the contract particulars.

4.20   If the contractor fails to complete the works or any section by the relevant completion date, liquidated damages become payable (see Figure 4.2).

Pre-agreed adjustment

4.21   There are two processes that can result in a pre-agreed adjustment to the completion date (under clause 2.26.2 'Pre-agreed Adjustment' is defined as 'the fixing of a revised Completion Date for the Works or a Section by the Confirmed Acceptance of a Variation Quotation or an Acceleration Quotation'). Both of these processes are set out in Schedule 2 to the form. The first is where the employer operates the 'Variation Quotation' mechanism set out in paragraph 1.2 of the Schedule, whereby the contractor is required to estimate not only the addition to the contract sum, but also any extension of time appropriate for the proposed variation. If the employer and contractor can subsequently agree on an amount, then this is binding on the parties (see paragraph 6.14).

4.22   The second process is through the confirmed acceptance of an acceleration quotation (Schedule 2:2). This provision allows the employer to investigate the possibility of achieving practical completion before the completion date. If invited by the employer, the contractor must provide, within 21 days, a quotation identifying the time that can be saved and the required increase in the contract sum, or explain why any acceleration is impracticable. The employer must accept or reject the proposal within seven days, and a confirmed acceptance would be binding on the parties. If the employer does not accept the quotation, the contractor is paid a reasonable sum for the cost of its preparation.

Extensions of time

Principle

4.23   An important reason for an extension of time clause is to preserve the employer's right to liquidated damages, in the event that the contractor fails to complete on time due in part to some action for which the employer is responsible. If there were no provisions to grant extensions of time, and a delay occurred that was caused at least in part by the employer, this would in effect be a breach of contract by the employer and the contractor would no longer be bound to complete by the completion date (*Peak Construction* v *McKinney Foundations*). The employer would therefore lose the right to liquidated damages, even though some of the blame for the delay may rest with the contractor. On the same principle, where the contract does provide for extending time, but these provisions are not operated, the employer will not be able to levy damages where it is in part responsible for the delay. The phrase 'time at large' is often used to describe these situations. In most cases, however, the contractor would remain under an obligation to complete within a reasonable time.

Peak Construction (Liverpool) Ltd v *McKinney Foundations Ltd* (1970) 1 BLR 111 (CA)

Peak Construction was the main contractor on a project to construct a multi-storey block of flats for Liverpool Corporation. The main contract was not on any of the standard forms, but was drawn up by the Corporation. McKinney Foundations Ltd was the sub-contractor nominated to design and construct the piling. After the piling was complete and the sub-contractor had left the site, serious defects were discovered in one of the piles and, following further investigation, minor defects were found in several other piles. Work was halted while the best strategy for remedial work was debated between the parties. The city surveyor did not accept the initial remedial proposals, and it was agreed that an independent engineer would prepare an alternative proposal. The Corporation refused to agree to accept his decision in advance, and delayed making the appointment. Altogether it was 58 weeks before work resumed (although the remedial work took only six weeks) and the main contractor brought a claim against the sub-contractor for damages. The Official Referee, at first instance, found that the entire 58 weeks constituted delay caused by the nominated sub-contractor and awarded £40,000 damages for breach of contract, based in part on liquidated damages which the Corporation had claimed from the contractor. McKinney appealed, and the Court of Appeal found that the 58-week delay could not possibly entirely be due to the sub-contractor's breach, but was in part caused by the tardiness of the Corporation. This being the case, and as there were no provisions in the contract for extending time for delay on the part of the Corporation, it lost its right to claim liquidated damages, and this component of the damages awarded against the sub-contractor was disallowed. Even if the contract had contained such a provision, the failure of the architect to exercise it would have prevented the Corporation from claiming liquidated damages. The only remedy would have been for the Corporation to prove what damages it had suffered as a result of the breach.

Procedure

4.24   In SBC16 the provisions for granting an extension of time are set out under clauses 2.26–2.29. The contractor gives written notice 'forthwith' to the contract administrator, when progress to the works or any section is being or is likely to be delayed (cl 2.27.1). The use of the terms 'forthwith' and 'is likely' suggests that the notice should be given as soon as a potential problem becomes apparent, without waiting to see whether it results in a measurable delay. The notice must be given whether or not completion is likely to be delayed, and whether or not the delay is caused by a 'Relevant Event' (i.e. the requirement to give a notice is not limited to circumstances where the contractor is claiming an extension of time).

4.25   The notice should set out the material circumstances and causes of the delay and identify any relevant events (cl 2.27.1). The notice must include or be followed by further particulars in respect of each and every relevant event, including the delay caused by each of those events and an estimate by the contractor of its effect on completion (cl 2.27.2). The contractor's notice and particulars, including its estimate, appear to be a condition precedent for the granting of an extension of time (*Multiplex Constructions (UK) Limited v Honeywell Control Systems Limited*). The contract administrator is therefore not obliged to issue an extension until the contractor has properly complied with clause 2.27. The contractor is obliged to keep the contract administrator informed of any changes in the estimated delay and in the particulars provided, and to 'supply such further information as the Architect/Contract Administrator may at any time reasonably require' (cl 2.27.3). It is suggested that the contract administrator's right to require further information is not restricted to any changes in the delay, but would include information relating to any of the matters raised in the notice.

> *Multiplex Constructions (UK) Limited v Honeywell Control Systems Limited* (No. 2) [2007] EWHC 447
>
> Wembley National Stadium Limited (WNSL) contracted with Multiplex Constructions (UK) Ltd (Multiplex), to construct the new Wembley National Stadium. Multiplex engaged Honeywell Control Systems Limited (Honeywell) under a sub-contract to design, supply and install various electronic systems. By the time Honeywell entered into the sub-contract, substantial delays to the project had already occurred. Multiplex issued three revised programmes to Honeywell, extending the completion date to 31 March 2006. The date passed without completion being achieved. No further programmes were issued by Multiplex. Honeywell maintained that the issue of the three programmes entitled it to claim prolongation costs and other financial relief. Honeywell argued that time had been set at large, due to its non-compliance with the conditions precedent. The judge decided that the contractor still had to comply with any notice provisions before an extension of time application could be entertained, even in relation to acts of prevention by the employer: 'Contractual terms requiring a contractor to give prompt notice of delay serve a valuable purpose; such notice enables matters to be investigated while they are still current'.

4.26   The amount of detail required is sometimes disputed, but in order to give proper consideration to a notice there must be sufficient information. It is likely that clause 2.27.2 would be interpreted as an obligation to supply a reasonable level of detail, particularly in respect of information to which only the contractor has access. The contract administrator should ask for more information if this is necessary to make a fair and reasonable assessment. This must never be regarded as a delaying tactic, however. The contract administrator must reach a decision 'on the balance of probability', taking into account all information available, including matters of which the administrator has knowledge and matters that can be established through reasonable enquiry. An extension which later appears insufficient in the light of further information can always be adjusted at review.

4.27   In regard to relevant events, the following points should be noted:

● The contractor will be entitled to an extension if compliance with cl 3.22 (antiquities) is necessary.

- The contractor will be entitled to an extension following any exercise of its right of suspension arising from non-payment of amounts due by the employer (cl 2.29.6).

- 'Any impediment, prevention or default … ' (cl 2.29.7) covers a very wide range of possible acts by the employer, contract administrator, quantity surveyor or any 'Employer's Person', and would include, for example, failure to provide a drawing at the time shown on the information release schedule.

- Clause 2.29.8 applies only to statutory undertakers operating independently. Work carried out by a statutory undertaker under a contract with the employer would fall under clause 2.29.7, and if under a contract with the contractor, any delays would be at the contractor's risk.

- 'Weather' is to be exceptional and adverse (i.e. not that which would be expected at the time of year in question) (cl 2.29.9). The effect of the weather is assessed at the time the work is actually carried out, not when it should have been, according to the contractor's programme. The contractor will normally provide weather records to support its claim.

- 'Specified Perils' (cl 2.29.10) can, under certain circumstances, include events caused by the contractor's own negligence.

- Civil commotion and terrorism under 2.29.11 includes the threat of terrorism, and activities of local authorities in dealing with such threats.

- There is very wide protection afforded with respect to strikes, and not simply those directly affecting the works, but also those causing difficulties in preparation and transportation of goods and materials, or the preparation of the design for the contractor's designed portion. Such strikes will not necessarily be confined to the UK and, given the current extent of overseas imports, the effects could be considerable (cl 2.29.12).

- SB16 includes two new relevant events: instructions where the contractor makes reasonable objection to a named specialist (cl 2.29.2.3), and a named specialist becoming insolvent (cl 2.29.14, see paragraph 5.70).

- 'Force majeure' (cl 2.29.15) is a French term used 'with reference to all circumstances independent of the will of man, and which it is not in his power to control'. It includes Acts of God and other matters outside the control of the parties. However, many items under this category, for example strikes, fire and weather, are dealt with elsewhere in the contract.

4.28   The contract administrator must respond 'as soon as is reasonably practicable' and within 12 weeks from receipt of the 'required particulars' (cl 2.28.2) (see Figure 4.3). It is not clear whether this phrase is referring to the particulars given in the contractor's notice or subsequently (cl 2.27.2), or any particulars requested by the contract administrator (cl 2.27.3). It is likely, however, that time would start to run from the point where sufficient information has been provided, and if this occurred with the first notice then time would run from that date. In any event, the requirement to respond 'as soon as is reasonably practicable' should not be overlooked. What time period is 'reasonably practicable' will depend on the nature of the delay, and the complexity of the information to be assessed, but once sufficient information is available the contract administrator should proceed to making a decision, and not wait until the end of the 12-week period. If the completion date is less than 12 weeks away the contract administrator must endeavour to respond by the completion date. This is not stated to be an absolute obligation, but as discussed above the contract administrator should proceed as soon as is practicable – any undue delay

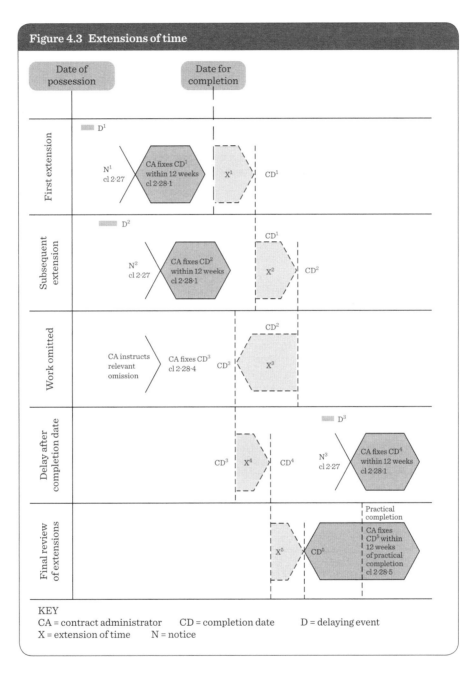

Figure 4.3 Extensions of time

KEY
CA = contract administrator   CD = completion date   D = delaying event
X = extension of time   N = notice

may result in the contractor claiming that it is no longer bound to finish by the current completion date.

4.29   The contract administrator must either fix a new completion date for the works or section, or notify the contractor that no extension of time is due (cl 2.28.2). It is important to note that SBC16 requires the contract administrator to state in the decision the amount of extension attributed to each relevant event, in other words the contract administrator must apportion the decision (cl 2.28.3.1).

4.30   The contract administrator may reduce a previous extension of time by fixing an earlier completion date having regard to 'Relevant Omissions', i.e. where work has been omitted through a variation instruction (cl 2.28.4), in which case the contractor must be notified of the reduction attributed to each relevant omission (cl 2.28.3.2). An extension or reduction may also result from a pre-agreed adjustment. The contract administrator may not, however, fix a date earlier than the date for completion entered in the contract particulars (cl 2.28.6.3), nor alter the length of a pre-agreed adjustment unless the work which was the subject of the pre-agreed adjustment is itself subject to a relevant omission (cl 2.28.6.4).

4.31   It appears that the contract administrator may not issue an extension of time before the completion date unless a notice of delay under clause 2.27 has been given by the contractor, and may not award one except in respect of relevant events identified by the contractor. If the contract administrator becomes aware of such matters, he or she could alert the contractor, but would be under no obligation to do so.

4.32   After the completion date of the works or any section has passed, the contract administrator may review the extensions of time given, and must do so prior to 12 weeks after practical completion (cl 2.28.5). The review may extend or bring forward the completion date, or confirm the date previously fixed and must be notified to the contractor together with the apportionment required by clause 2.28.3. At this point the contract administrator must take into account relevant events not notified by the contractor (cl 2.28.5.1). As above, the contract administrator can only reduce the extensions of time already given if a reduction or omission of work instructed after the last completion date was fixed would justify this (cl 2.28.5.2), and may not fix a date earlier than the original 'Date for Completion'.

4.33   If the contract administrator awards a further extension of time in respect of relevant events which occur after the date for completion or any extended completion date, i.e. when the contractor is in 'culpable delay' (*Balfour Beatty* v *Chestermont Properties*), the extension is added onto the date that has passed, referred to as the 'net' method of extension. It should be noted that this only operates under clause 2.28.5, discussed above. Contractors' applications under clause 2.27.1 can only relate to relevant events that occur before the practical completion date (implied by the wording of clause 2.28.1.2 'is likely to be').

Balfour Beatty Building Ltd v *Chestermont Properties Ltd* (1993) 62 BLR 1

In a contract on JCT80 the works were not completed by the revised completion date and the architect issued a non-completion certificate. The architect then issued a series of variation instructions and a further extension of time, which had the effect of fixing a completion date two-and-a-half months before the first of the variation instructions. He then issued a further non-completion certificate and the employer proceeded to deduct liquidated damages. The contractor took the matter to arbitration, and then appealed certain decisions on preliminary questions given by the arbitrator. The court held that the architect's power to grant an extension of time pursuant to clause 25.3.1.1 (equivalent to SBC16 cl 2.21.1) could only operate in respect of relevant events that occurred before the original or the previously fixed completion date, but the power to grant an extension under clause 25.3.3 (equivalent to SBC16 cl 2.28.5) applied to any relevant event. The architect was right to add the extension of time retrospectively (termed the 'net' method).

Assessment

4.34   It is an obligation on the contract administrator to issue extensions of time when properly due and any failure on the part of the contract administrator to do so is a breach on the part of the employer. The contract administrator has no power to grant extensions of time except for the relevant events. In every case the contract administrator should assess the effect of the delay on the contract completion date. The contractor's programme can be used as a guide but is not binding. The effect on progress is assessed in relation to the work being carried out at the time of the delaying event, rather than the work that was programmed to be carried out.

4.35   Clause 2.28.6.1 contains the important proviso that the contractor must 'constantly use his best endeavours to prevent delay'; therefore, the contract administrator can assume that the contractor will take steps to minimise the effect of the delay on the completion date. The phrase 'best endeavours' appears to suggest something more than 'reasonable' or 'practicable', but it is unlikely to extend to excessive expenditure. It should also be noted that the contractor is not to be given an extension of time where delays are caused by the defaults set out in clause 2.20, i.e. errors in the contractor's proposals, or delays in submitting contractor's design documents.

4.36   The effects of any delay on completion – taking into account the contractor's 'best endeavours' – are not always easy to predict. The contract administrator is required to reach an opinion, and in doing this the contract administrator owes a duty to both parties to be fair and reasonable (*Sutcliffe* v *Thackrah*, see paragraph 7.2). This applies even where the delay has been caused by the contract administrator; for example, where the contract administrator has failed to issue drawings within the time limits stipulated in the contract.

4.37   It sometimes happens that two or more delaying events can occur simultaneously, or with some overlap, and this can raise difficult questions with respect to the awarding of extensions of time. In the case of concurrent delays involving two or more relevant events, it has been customary to grant the extension in respect of the dominant reason. However, this approach was considered inappropriate in relation to direct loss and/or expense (*H Fairweather & Co.* v *Wandsworth*). The provisions in clause 2.28.3 make it clear that the contract administrator must apportion the total delay between the various contributing causes.

> *H Fairweather & Co. Ltd* v *London Borough of Wandsworth* (1987) 39 BLR 106
>
> Fairweather entered into a contract with the London Borough of Wandsworth to erect 478 dwellings. The contract was on JCT63. Pipe Conduits Ltd was the nominated sub-contractor for underground heating works. Disputes arose and an arbitrator was appointed who made an interim award. Leave to appeal was given on several questions of law arising out of the award. The arbitrator had found that where a delay occurred which could be ascribed to more than one event, the extension should be granted for the dominant reason. Strikes were the dominant reason, and the arbitrator had therefore granted an extension of 81 weeks for this reason, and made it clear that this reason did not carry any right to direct loss and/or expense. The court stated that an extension of time was not a condition precedent to an award of direct loss and/or expense, and that the contractor would be entitled to direct loss and/or expense for other events which had contributed to the delay.

4.38   Where one overlapping delaying event is a relevant event and the other is not (in other words, one is the employer's risk and the other the contractor's), a difficult question arises as to the extension of time due. The instinctive reaction of many assessors might be to 'split the difference', given that both parties have contributed to the delay. However, the more logical approach is that the contractor should be given an extension of time for the full length of delay caused by the relevant event, irrespective of the fact that during the overlap the contractor was also causing delay. Taking any other approach – for example, splitting the overlap period and awarding only half of the extension to the contractor – could result in the contractor being subject to liquidated damages for a delay partly caused by the employer. The courts have normally adopted this analysis (*Henry Boot Construction (UK) Limited* v *Malmaison Hotel*). More recently, in the much publicised Scottish case of *City Inn Ltd* v *Shepherd Construction Ltd*, the court stated that a proportional approach would be fairer. This decision, however, is not binding on English courts, and has not been followed (see *Walter Lilly & Co. Ltd* v *Giles Mackay & DMW Ltd*); therefore the *Malmaison* approach remains the correct one to adopt.

Henry Boot Construction (UK) Ltd v *Malmaison Hotel (Manchester) Ltd* (1999) 70 Con LR 32 (TCC)

The employer, Malmaison, engaged Henry Boot to construct a new hotel in Piccadilly, Manchester. Completion was fixed for 21 November 1997, but was not achieved until 13 March 1998. However, extensions of time were issued by the architect revising the date for completion to 6 January 1998. Malmaison deducted liquidated damages from the contract sum. Although Henry Boot claimed further extensions of time in respect of a number of alleged relevant events, no further extensions of time were awarded. The case went to arbitration and the decision was challenged through court proceedings. Among other matters, the judge considered concurrency. If it can be shown that there are two equal and concurrent causes of delay, for which the employer and contractor are respectively responsible, then the contractor is still entitled to an extension of time. Judge Dyson illustrated his views on concurrency by citing the example of the start of a project being held up for one week by exceptionally inclement weather (a 'Relevant Event'), while at the same time the contractor suffered a shortage of labour (not a 'Relevant Event'). In effect, the two delays and causes were concurrent. In this situation, Judge Dyson said that the contractor should be awarded an extension of time of one week, and an architect should not deny the contractor an extension on the grounds that the project would have been delayed by the labour shortage.

City Inn Ltd v *Shepherd Construction Ltd* [2008] CILL 2537 Outer House Court of Session

In considering a case involving a dispute over extensions of time under a JCT80 form of contract, the court considered earlier authorities and the principles underlying extension of time clauses and set out several propositions. These included that where there are several causes of delay and where a dominant cause can be identified, the assessor can use the dominant cause and set aside immaterial causes. However, where there are two causes of delay, only one of which is a contractor default, the assessor may apportion delay between the two events. The CILL editors describe this as going 'further than any recent authority' on concurrency. Assessors should note that this is a Scottish case which has not, to date, been followed in English courts.

Walter Lilly & Co. Ltd v *Giles Mackay & DMW Ltd* [2012] EWHC 649 (TCC)

This case concerned a contract to build Mr and Mrs Mackay's, and two other families', luxury new homes in South Kensington, London. The contract was entered into in 2004 on the JCT Standard Form of Building Contract 1998 Edition with a Contractor's Designed Portion Supplement. The total contract sum was £15.3 million, the date for completion was 23 January 2006 and liquidated damages were set at £6,400 per day. Practical completion was certified in August 2008. The contractor (Walter Lilly) issued 234 notices of delay and requests for extensions of time, of which fewer than a quarter were answered. The contractor brought a claim for, among other things, an additional extension of time. The court awarded a full extension up to the date of practical completion. It took the opportunity to review approaches to dealing with concurrent delay, including that in the case of *Henry Boot Construction (UK) Ltd* v *Malmaison Hotel (Manchester) Ltd* (which established that the contractor is entitled to a full extension of time for delay caused by two or more events, provided one is an event which entitles it to an extension under the contract), and the alternative approach in the Scottish case of *City Inn Ltd* v *Shepherd Construction Ltd* (where the delay is apportioned between the events). The court decided that the former was the correct approach in this case. As part of its reasoning, the court noted that there was nothing in the relevant clauses to suggest that the extension of time should be reduced if the contractor was partly to blame for the delay.

Partial possession

4.39   Possession by the employer of completed parts of the works (or any section) ahead of practical completion of the whole works is provided for under clauses 2.33–2.37. This 'partial possession' requires the agreement of the contractor, which cannot be unreasonably withheld (cl 2.33 and 1.11). The contract administrator must issue a notice to the contractor identifying precisely the extent of the 'Relevant Part' and the date of possession (the 'Relevant Date'). This statement should be prepared with great care, if necessary using a drawing to illustrate the extent. The insurers should be notified where relevant.

4.40   The statement must be issued immediately after the part is taken into possession but, in practice, it would be wise to circulate the drawings and information in advance, so that the details of what will occur are clear to all parties. The partial possession may affect other operations on site, in which case it could constitute an 'impediment' (e.g. if access to other parts is affected and causes disruption to the programme), therefore it would be wise to clarify the position with the contractor before a decision is made, and the employer should consider all possible contractual consequences in terms of delay and claims for loss and/or expense.

4.41   Practical completion is 'deemed to have occurred' for the 'Relevant Part' of the works on the relevant date (cl 2.34). The rectification period for that part is deemed to have commenced (cl 2.34) and the certificate of making good defects has to be issued for that part separately (cl 2.35). However, it would appear that this remains part of the works and is still to be included under the practical completion certificate.

4.42   Liquidated damages are reduced by the proportion of the value of the possessed part of the works to the contract sum (cl 2.37). The effect of clause 4.19.2.2 is that half of the retention is released for that proportion of the works. If Insurance Option A, B or C2 applies, the employer may wish to consider insuring the part as the contractor's obligation to insure the works will cease (cl 2.36).

4.43   It is important to note that the fact that significant work remains outstanding has not prevented the courts from finding that 'partial possession' has been taken of the whole works, in situations where a tenant has effectively occupied the whole building, allowing access to the contractor for remedial work (see *Skanska Construction (Regions) Ltd v Anglo-Amsterdam Corporation Ltd*). If the employer wishes to occupy the building, but parties do not intend clause 2.33 to take effect for the whole project, they must make it clear, under a carefully worded agreement, what the contractual consequences of the occupation are intended to be (see below).

> *Skanska Construction (Regions) Ltd* v *Anglo-Amsterdam Corporation Ltd* (2002) 84 Con LR 100
>
> Anglo-Amsterdam Corporation (AA) engaged Skanksa Construction (Skanska) to construct a purpose-built office facility under a JCT81 With Contractor's Design form of contract. Clause 16 had been amended to state that practical completion would not be certified unless the certifier was satisfied that any unfinished works were 'very minimal and of a minor nature and not fundamental to the beneficial occupation of the building'. Clause 17 of the form stated that practical completion would be deemed to have occurred on the date the employer took possession of 'any part or parts of the Works'.
>
> AA wrote to Skanska confirming that the proposed tenant for the building would commence fitting-out works on the completion date. However, the air-conditioning system was not functioning and Skanska had failed to produce operating and maintenance manuals. Following this date the tenant took over responsibility for security and insurance, and Skanska was allowed access to complete outstanding work. AA alleged that Skanska was late in the completion of the works and applied liquidated damages at the rate of £20,000 per week for a period of approximately nine weeks. Skanska argued that the building had achieved practical completion on time or that, alternatively, partial possession of the works had taken place and that, consequently, its liability to pay liquidated damages had ceased under clause 17.
>
> The case went to arbitration and Skanska appealed. The court was also unhappy with the decision and found that clause 17.1 could also operate when possession had been taken of all parts of the works and was not limited to possession of only part or some parts of the works. Accordingly, it found that partial possession of the entirety of the works had, in fact, been taken some two months earlier than the date of practical completion, when AA agreed to the tenant commencing fit-out works. Consequently, even though significant works remained outstanding, Skanska was entitled to repayment of the liquidated damages that had already been deducted by AA.

Use or occupation before practical completion

4.44   When the contractor is still in possession of the site, and the employer wishes to 'use or occupy the site or the Works or part of them', then clause 2.6 provides for this. The purposes for which the employer might require this are described simply as 'storage or otherwise', so, in theory at least, the clause places no limits on what form the use might take. The consent of the contractor in writing is required, and the contractor would not be unreasonable in withholding permission unless the intervention and inconvenience are likely to be minimal, or unless an agreement is reached over granting an extension of time and/or compensating the contractor for any losses. Before the contractor is required to give consent, the employer or the contractor as appropriate must notify the insurers (cl 2.6.1). If under Insurance Option A any additional premium is required, the contractor notifies the employer of the amount and this is added to the contract sum (cl 2.6.2).

Figure 4.4 'Practice' section, *RIBA Journal* (February 1992)

Employer's possession before practical completion under JCT contracts

It is not uncommon for the employer, after the completion date has passed, to wish to take possession of the Works before the contractor has achieved practical completion. In this event an ad hoc agreement between employer and contractor is required to deal with the situation. In respect of such an agreement, members may wish to have regard to the following note . . .

Outstanding items

Where it is known to the architect that there are outstanding items, practical completion should not be certified without specially agreed arrangements between the employer and the contractor. For example, in the case of a contract where the contract completion date has passed it could be so agreed that the incomplete building will be taken over for occupation, subject to postponing the release of retention and the beginning of the defects liability period until the outstanding items referred to in a list to be prepared by the architect are completed, but relieving the contractor from liability for liquidated damages for delay as from the date of occupation, and making any necessary changes in the insurance arrangements. In such circumstances either the Certificate of Practical Completion form should not be used or it should be altered to state or refer to the specially agreed arrangements. In making such arrangements the architect should have the authority of the client-employer.

When the employer is pressing for premature practical completion there is a need to be particularly careful where there are others who are entitled to rely on the issue of a Practical Completion Certificate and its consequences. In the case where part only of the Works is ready for hand-over the partial possession provisions can be operated to enable the employer with the consent of the contractor to take possession of the completed part.

4.45   Towards the end of a project the situation often arises where the contractor has not completed by the date for completion and, although no sections of the works are sufficiently complete to allow the employer to take possession of those parts under clauses 2.33–2.37, the employer is nevertheless anxious to occupy at least part of the works. Clause 2.6 appears to cover early occupation, but great care must be taken to agree to all the contractual consequences before the occupation takes place. A suggestion was put forward in the 'Practice' section of the *RIBA Journal* (February 1992), which has frequently proved useful in practice (see Figure 4.4). In this arrangement, in return for being allowed to occupy the premises, the employer agrees not to claim liquidated damages during the period of occupation. Practical completion obviously cannot be certified, and there is no release of retention money until it is. Matters of insuring the works will need to be settled with the insurers.

4.46   Because such an arrangement would be outside the terms of the contract it should be covered by a properly drafted agreement which is signed by both parties. (The cases of *Skanska* v *Anglo-Amsterdam Corporation* above and *Impresa Castelli* v *Cola* illustrate the importance of drafting a clear agreement.) It may also be sensible to agree that in the event that the contractor still fails to achieve practical completion by the end of an agreed period, liquidated damages would run again, possibly at a reduced rate. In most

Impresa Castelli SpA v *Cola Holdings Ltd* (2002) CLJ 45

Impresa agreed to build a large four-star hotel for Cola Holdings Ltd (Cola), using the JCT Standard Form of Building Contract With Contractor's Design, 1981 edition. The contract provided that the works would be complete within 19 months from the date of possession. As the work progressed it became clear that the date for completion of February 1999 was not going to be met, and the parties agreed a new completion date in May 1999 (with the bedrooms available to Cola in March) and a new liquidated damages provision of £10,000 per day, as opposed to the original rate of £5,000. Once the agreement was in place, further difficulties with progress were encountered, which meant that the May 1999 completion date was also unachievable. The parties entered into a second variation agreement, which recorded that Cola would be allowed access to parts of the hotel to enable it to be fully operational by September 1999, despite certain works being incomplete (including the air conditioning). In September 1999, parts of the hotel were handed over, but Cola claimed that such parts were not properly completed. A third variation agreement was put in place with a new date for practical completion and for the imposition of liquidated damages.

Disputes arose and, among other matters, Cola claimed for an entitlement for liquidated damages. Impresa argued that it had achieved partial possession of the greater part of the works, therefore a reduced rate of liquidated damages per day was due. The court found that although each variation agreement could have used the words 'partial possession', they had in fact instead used the word 'access'. The court had to consider whether partial possession had occurred under clause 17.1 of the contract, which provides for deemed practical completion when partial possession is taken, or whether Cola's presence was merely 'use or occupation' under clause 23.3.2 of the contract. The court could find nothing in the variation agreements to suggest that partial possession had occurred. It therefore ruled that what had occurred related to use and occupation, as referred to in clause 23.3.2 of the contract, and the agreed liquidated damages provision was therefore enforceable.

circumstances this arrangement would be of benefit to both parties, and is far preferable to issuing a heavily qualified practical completion certificate listing 'except for' items.

Practical completion

4.47   The contract administrator is obliged to certify practical completion of the works or a section (cl 2.30) when, in the contract administrator's opinion, the following criteria are fulfilled:

● practical completion of the works or the section is achieved (see below);

● the contractor has complied 'sufficiently' with clauses 2.40 and 3.23 in respect of the supply of documents and information (i.e. as-built drawings and information required for the health and safety file).

4.48   The wording of clause 2.30 is not as clear as it might be, in that the first part appears to be stating that the contract administrator is certifying three separate things: 'practical completion of the Works or a Section' plus compliance with two other clauses. However, it must be remembered that this wording is the result of later insertions being placed into a clause which originally referred solely to practical completion of the works. It is suggested that a correct analysis of this clause is that practical completion only occurs when all three conditions are met, the principal argument for this being that only one date is entered on the certificate. This should be the date when the last condition is fulfilled; in other words, if there is a delay before receiving the as-built drawings, the date of their receipt should be the date on the certificate, irrespective of the fact that the works were complete days or even weeks earlier.

4.49   It should be noted that the use of the term 'complied sufficiently' may allow the contract administrator to use its discretion in issuing the certificate with some very minor information missing. The contract administrator should, however, be very careful not to place the employer in a position where it would be in breach of the CDM Regulations.

4.50   Deciding when the works have reached practical completion often causes some difficulty. In the leading commentary *Keating on Construction Contracts*[1], the editors submitted that the following is the correct analysis:

(a)   the works can be practically complete notwithstanding that there are latent defects;

(b)   a practical completion certificate may not be issued if there are patent defects. The rectification period is provided in order to enable defects not apparent at the date of practical completion to be remedied (*City of Westminster* v *Jarvis & Sons Ltd* and *H W Nevill (Sunblest)* v *William Press*);

(c)   practical completion means the completion of all the construction that has to be done (*City of Westminster* v *Jarvis & Sons Ltd*);

(d)   however, the contract administrator is given a discretion under clause 2.30 to certify practical completion where there are very minor items of work left incomplete, on *de minimis* principles (*H W Nevill (Sunblest)* v *William Press*).

City of Westminster v *Jarvis & Sons Ltd* (1970) 7 BLR 64 (HL)

Jarvis entered into a contract with the City of Westminster to construct a multi-storey car park in Rochester Row. The contract was on JCT63, which included 'delay on the part of a nominated sub-contractor ... which the Contractor has taken all reasonable steps to avoid' (cl 23(g)). The piling was carried out by the nominated sub-contractor, which completed its work by the date required under its sub-contract and withdrew from site. Subsequently, defects were discovered in many of the piles and the remedial works caused a delay of over 21 weeks to the contractor. The House of Lords found that on a proper interpretation of clause 23(g), delay on the part of the nominated sub-contractor only occurred if it had failed to complete its work by the date in the sub-contract. The clause did not apply after the works had been accepted as complete. In addressing the question of whether there was a delay in completion the court also had to consider what was meant by 'practical completion' of the sub-contract works. Viscount Dilhourne stated: 'The contract does not define what is meant by "practically completed". One would normally say that a task was practically completed when it was almost but not entirely finished; but 'practical completion' suggests that that is not the intended meaning and that what is meant is the completion of all the construction work that has to be done' (at page 75).

---

[1]   Stephen Furst and Vivien Ramsey (eds), *Keating on Construction Contracts*, 10th edn (London: Sweet & Maxwell, 2016)

> *H W Nevill (Sunblest) Ltd* v *William Press & Son Ltd* (1981) 20 BLR 78
>
> William Press entered into a contract with Sunblest to carry out foundations, groundworks and drainage for a new bakery on a JCT63 contract. A practical completion certificate was issued, and new contractors commenced a separate contract to construct the bakery. A certificate of making good defects and a final certificate were then issued for the first contract, following which it was discovered that the drains and the hard standing were defective. William Press returned to site and remedied the defects, but the second contract was delayed by four weeks and Sunblest suffered damages as a result. It commenced proceedings, claiming that William Press was in breach of contract and in its defence William Press argued that the plaintiff was precluded from bringing the claim by the conclusive effect of the final certificate. Judge Newey decided that the final certificate did not act as a bar to claims for consequential loss. In reaching the decision he considered the meaning and effect of the practical completion certificate and stated: 'I think that the word "practically" in clause 15(1) gave the architect a discretion to certify that William Press had fulfilled its obligation under clause 21(1) where very minor *de minimis* work had not been carried out, but that if there were any patent defects in what William Press had done then the architect could not have issued a certificate of practical completion' (at page 87).

4.51   In *Keating on Construction Contracts* it is also advised that the discretion of the contract administrator should be exercised with caution, and this is undoubtedly sound advice. The decision as to when practical completion has occurred is one of the most critical decisions that the contract administrator has to make in administering the contract, as the consequences to the employer are significant. The author frequently encounters instances of contract administrators certifying practical completion qualified by long lists of 'snagging' items. Even though the employer, anxious to move into the newly completed works, may initially agree to the early certification, the following contractual problems will remain unresolved:

- half of the retention will be released for that section or the whole works, leaving only 1.5 per cent retention in hand (cl 4.19.2). This puts the employer at considerable risk, as the 1.5 per cent is only intended to cover latent, not patent, defects;

- the relevant rectification period begins (cl 2.38);

- the onus shifts to the contract administrator to instruct all necessary outstanding work under clause 2.38.1. If the contract administrator fails to instruct something, the contractor would have no authority to enter the site to carry it out – therefore, the contract administrator will inevitably become involved in managing and programming the outstanding work;

- the contractor's liability for negligent damage to the works commences (cl 6.3.4);

- possession of the site passes to the employer and, depending on the insurance arrangements, the contractor might no longer cover the insurance of the works. The insurers will need to be informed about the programme for the outstanding works;

- the contractor's liability for liquidated damages ends (cl 2.32.2);

- the employer will be the 'occupier' for the purposes of the Occupiers Liability Acts 1957 and 1984 and also may be subject to claims regarding health and safety.

4.52   A certificate must be issued as soon as the criteria in clause 2.30 are met for each section (termed a 'Section Completion Certificate') and for the whole works (the 'Practical Completion Certificate'). It is sufficient that the contractor complies with the technical standards and requirements set out in the contract; it does not have to satisfy any other unstated general employer requirements (*Laing O'Rourke* v *Healthcare Support*). Once practical completion is certified the employer is obliged to accept the works. Employers who have additional requirements or who may wish to accept the works only on the date in the contract would need to amend the wording.

> *Laing O'Rourke Construction Ltd (formerly Laing O'Rourke Northern Ltd)* v *Healthcare Support (Newcastle) Ltd* [2014] EWHC 2595 (TCC)
>
> This case concerned a hospital project, where the relevant contract provided for the engagement of an independent tester who would certify practical completion of the claimant's works in accordance with the completion criteria specified in the project agreement. The independent tester refused to certify practical completion due to the employer's complaints about the quality and conformity of certain aspects of the works (toilet area size, daylight levels, window restrictors, link bridge steelwork and room temperature). The contractor argued that these matters fell outside the completion criteria, and that compliance with the criteria was sufficient. The court agreed with this view. However, in reaching its opinion it acknowledged that, where there are departures from the criteria, the tester would nevertheless be justified in issuing the certificate where the departure would not have any material adverse impact on the ability of the employer to enjoy and use the buildings for the purposes anticipated by the contract. There was no justification to imply a term that any breach of the contractual requirements, however technical or minor, would prevent certification of practical completion.

4.53   If the employer wishes to move in before the works are complete, rather than issuing a practical completion certificate prematurely it would be better to arrange for partial possession or for a special agreement to be reached, as set out above in paragraphs 4.44–4.46. It should be noted that simply because the employer moves in, this does not necessarily mean that 'practical completion' has been achieved, nor that liquidated damages would no longer be applicable (e.g. see *BFI Group of Companies Ltd* v *DCB Integration Systems Ltd*, at paragraph 4.58), although should the occupation interfere with the carrying out of the works this may give rise to claims.

Procedure at practical completion

4.54   The contract sets out no procedure for what happens at practical completion, it simply requires the contract administrator to certify it. The contract bills may set out a procedure, and the contract administrator should check carefully at tender stage to ensure that the procedure is satisfactory. In particular, the specification or bills may stipulate commissioning procedures for mechanical and electrical services, and testing procedures to demonstrate compliance with Building Regulations.

4.55   Leading up to practical completion it appears to be widespread practice for contract administrators to issue 'snagging' lists, sometimes in great detail and on a room-by-room basis. The contract does not require this, and neither do most standard terms of appointment. Under the contract, responsibility for quality control and snagging rests entirely with the contractor. In adopting this role the contract administrator may be assisting the contractor, and although this may appear to benefit the employer, it may lead to confusion over the liability position, which could cause problems at a future date.

4.56   It is common practice for the contractor to arrange a 'handover' meeting. The term is not used in SBC16, and although handover meetings can be of use, particularly in introducing the finished project to the employer, it may be better not to set out complex or inflexible procedures in the bills of quantities. Even where a handover meeting has been arranged, or the contractor has stated in writing that the works are complete, it remains the contract administrator's responsibility to decide when practical completion has been achieved. If the contract administrator feels that the works are not complete, there is no obligation to justify this opinion with schedules of outstanding items. It is suggested that the best course may be to draw attention to typical items, but to make it clear that the list is indicative and not comprehensive.

Failure to complete by the completion date

4.57   In the event of failure to complete a section or the works by the relevant date, the contract administrator is required to certify this fact by means of a 'Non-Completion Certificate' (cl 2.31). As this certificate is a condition precedent to deduction of liquidated damages (cl 2.32.1.1), it is important that it is issued promptly. Once the certificate has been issued, the contractor is said to be in 'culpable delay'. The employer, provided that it has issued the necessary notices (see paragraphs 4.60–4.62), may then deduct the damages from the next interim certificate, or reclaim the sum as a debt. Note that fluctuations provisions are frozen from this point. If a new completion date is later set, this has the effect of cancelling the non-completion certificate and no additional written cancellation is needed (cl 2.32.3).

Liquidated damages

4.58   The agreed rate for liquidated damages for each section or for the works is entered in the contract particulars. This is normally expressed as a specific sum per week (or other unit) of delay, to be allowed by the contractor in the event of failure to complete by the completion date (note there may be several different rates where the works are divided into sections). As a result of two decisions in the Supreme Court, it is no longer considered essential that the amount is calculated on the basis of a genuine pre-estimate of the loss likely to be suffered (*Cavendish Square Holdings* v *El Makdessi* and *ParkingEye Limited* v *Beavis*; see also *Alfred McAlpine Capital Projects* v *Tilebox*). Provided that the amount is not 'out of all proportion' to the likely losses, the damages will be recoverable without the need to prove the actual loss suffered, irrespective of whether the actual loss is significantly less or more than the recoverable sum (*BFI Group of Companies* v *DCB Integration Systems*). In other words, once the rate has been agreed, both parties are bound by it. Of course, for practical reasons, the rate should always be discussed with the employer before inclusion in the tender documents, and an amount that will provide adequate compensation included to cover, among other things, any additional professional fees that may be charged during this period. If 'nil' is inserted then this may preclude the employer from claiming any damages at all (*Temloc* v *Errill*), whereas if the contract particulars are left blank the employer may be able to claim general damages.

Cavendish Square Holdings v *El Makdessi* and *ParkingEye Limited* v *Beavis*, Supreme Court 2015

In this landmark case the Supreme Court restated the law regarding whether a liquidated damages clause may be considered a penalty. Key criteria for whether a provision will be penal are: if 'the sum stipulated for is extravagant and unconscionable in amount in comparison with the greatest loss that could conceivably be proved to have followed from the breach'; and whether the sum imposes a detriment on the contract breaker which is 'out of all proportion to any legitimate interest of the innocent party'. In determining these, the court must consider the wider commercial context.

Alfred McAlpine Capital Projects Ltd v *Tilebox Ltd* [2005] BLR 271

This case contains a useful summary of the law relating to the distinction between liquidated damages and penalties. A WCD98 contract contained a liquidated damages provision in the sum of £45,000 per week. On the facts, this was a genuine pre-estimate of loss and the actual loss suffered by the developer, Tilebox, was higher. The contractor therefore failed to obtain a declaration that the provision was a penalty. However, the judge also considered a different (hypothetical) interpretation of the facts whereby it was most unlikely, although just conceivable, that the total weekly loss would be as high as £45,000. In this situation also the judge considered that the provision would not constitute a penalty. In reaching this decision he took into account the facts that the amount of loss was difficult to predict, that the figure was a genuine attempt to estimate losses, that the figure was discussed at the time that the contract was formed and that the parties were, at that time, represented by lawyers.

BFI Group of Companies Ltd v *DCB Integration Systems Ltd* [1987] CILL 348

BFI employed DCB on the Agreement for Minor Building Works to refurbish and alter offices and workshops at its transport depot. BFI was given possession of the building on the extended date for completion, but two of the six vehicle bays could not be used for a further six weeks as the roller shutters had not yet been installed. Disputes arose which were taken to arbitration. The arbitrator found that the delay in completing the two bays did not cause BFI any loss of revenue, and that BFI was therefore not entitled to any of the liquidated damages. BFI was given leave to appeal to the High Court. HH Judge John Davies QC found that BFI was entitled to liquidated damages. It was quite irrelevant to consider whether in fact there was any loss. Liquidated damages do not run until possession is given to the employer but until practical completion is achieved, which may not be at the same time. Therefore, the fact that the employer had use of the building was also not relevant.

Temloc Ltd v *Errill Properties* (1987) 39 BLR 30 (CA)

Temloc entered into a contract with Errill Properties to construct a development near Plymouth. The contract was on JCT80 and was in the value of £840,000. '£ Nil' was entered in the contract particulars against clause 24.2, liquidated damages. Practical completion was certified around six weeks later than the revised date for completion. Temloc brought a claim against Errill Properties for non-payment of some certified amounts, and Errill counterclaimed for damages for late completion. It was held by the court that the effect of '£ nil' was not that the clause should be disregarded (because, for example, it indicated that it had not been possible to assess a rate in advance), but that it had been agreed that no damages would be payable in the event of late completion. Clause 24 is an exhaustive remedy and covers all losses normally attributable to a failure to complete on time. The defendant could not, therefore, fall back on the common law remedy of general damages for breach of contract.

4.59   Before liquidated damages may be claimed, the following preconditions must have been met:

- the contractor must have failed to complete the works by the completion date;
- the contract administrator must have fulfilled all duties with respect to the award of an extension of time;
- the contract administrator must have issued a non-completion certificate (cl 2.32.1.1);
- the employer must have informed the contractor before the date of the final certificate that it may require the payment of liquidated damages or deduct liquidated damages from monies due (cl 2.32.1.2).

4.60   If these preconditions are met then the employer may, not later than five days before the final date for payment of 'the amount payable under clause 4.26' (i.e. the final certificate), issue a notice in accordance with clause 2.32.2 (cl 2.32.1). It should be noted that, although clause 2.32.1 states the employer 'may' give this notice, in effect the notice *must* be issued if the employer wishes to claim liquidated damages.

4.61   Clause 2.32.1 therefore requires two types of notice: a general notice of intention that the employer 'may require' damages (cl 2.32.1.2) and a notice at the time a payment is required or a deduction is to be made (cl 2.32.1). This second notice requires more detail than the first; it must state that 'for the period between the Completion Date and the date of practical completion' the employer requires the contractor to pay the sum to the employer and/or intends to deduct liquidated damages from monies due, and whether the rate will be the contractual one or a lesser sum (cl 2.32.2). The requirement and notification must be reasonably clear, but there is no need for a great deal of detail (*Finnegan* v *Community Housing Association*).

> **J F Finnegan Ltd v Community Housing Association Ltd (1995) 77 BLR 22 (CA)**
>
> Finnegan Ltd was employed by the Housing Association to build 18 flats at Coram Street, West London, on the JCT SBC 1980. The contractor failed to complete the work on time and the contract administrator issued a certificate of non-completion. Following the practical completion certificate an interim certificate was issued. The employer sent a notice with the cheque honouring the certificate, which gave minimal information (i.e. not indicating how liquidated damages had been calculated). The Court of Appeal considered this sufficient to satisfy the requirement for the employer's written notice in clause 24.2.1 (similar to cl 2.32.2 in SBC16). Peter Gibson LJ stated (at page 33):
>
> > I consider that there are only two matters which must be contained in the written requirement. One is whether the employer is claiming a payment or a deduction in respect of LADs [liquidated and ascertained damages]. The other is whether the requirement relates to the whole or a part (and, if so, what part) of the sum for the LADs.
>
> He then stated (at page 35):
>
> > I would be reluctant to import into this commercial agreement technical requirements which may be desirable but which are not required by the language of the clause and are not absolutely necessary.
>
> The requirements relating to notices have now changed. However, there appears to be no reason why the general comments would not still apply, i.e. that the amount of information required would be no more than the minimum set out in the contractual provisions.

4.62   It may be possible for the clause 2.32.1.2 notice of intention and the clause 2.32.2 notice to be dealt with together, provided the document contains the necessary information and is issued at the right time. If the employer wishes to withhold or deduct all or any of the liquidated damages payable, footnote [38] to clause 2.32.2.2 explains that, in addition to the notice under clause 2.32.1, the employer must give the appropriate pay less notice under clause 4.11. This notice will explain the calculation for that particular deduction, i.e. the period over which the damages are claimed, and the rate applied. In effect, if a deduction from a certificate is intended, then all three notices will be needed prior to the first deduction being made; subsequent certificates will require only a pay less notice.

4.63   If an extension of time is given following the issue of a non-completion certificate then this has the effect of cancelling that certificate. A new non-completion certificate must be issued if the contractor then fails to complete by the new completion date (cl 2.31). The contract states that the employer does not need to inform the contractor again that it may claim damages, as clause 2.32.1.2 remains satisfied (cl 2.32.4). The contract does not address the question of whether a new clause 2.32.2 notice is required, but for the sake of clarity it may be wise to do this.

4.64   Once the completion date is adjusted the employer must, if necessary, repay any liquidated damages recovered for the period up to the new completion date (cl 2.32.3) and must do so within a reasonable period of time (*Reinwood v L Brown & Sons*). Clause 2.32.4 states that any notice which has previously been issued in accordance with clause 2.32.1.2 shall remain effective 'unless the Employer states otherwise in writing', notwithstanding that a further extension of time has been granted.

> *Reinwood Ltd* v *L Brown & Sons Ltd* [2007] BLR 305 (CA)
>
> This dispute concerned a contract on JCT98, with a date for completion of 18 October 2004, and liquidated damages at the rate of £13,000 per week. The project was delayed, and on 7 December 2005 the contractor made an application for an extension of time. On 14 December 2005, the contract administrator issued a certificate of non-completion under clause 24.1. On 11 January 2006, the contract administrator issued interim certificate no. 29 showing the net amount for payment as £187,988. The final date for payment was 25 January 2006.
>
> On 17 January 2006, the employer issued notices under clauses 24.2 and 30.1.1.3 of its intention to withhold £61,629 as liquidated damages, and the employer duly paid £126,359 on 20 January 2006.
>
> On 23 January 2006, the contract administrator granted an extension of time until 10 January 2006, following which the contractor wrote to the employer stating that the effect of the extension of time and revision of the completion date was that the employer was now entitled to withhold no more than £12,326. The amount due under interim certificate no. 29 was, therefore, £175,662. Subsequently, the contractor terminated the contract, relying partly on the late repayment of the balance by the employer.
>
> The appeal was conducted on the issue of whether the cancellation of the certificate of non-completion by the grant of an extension of time meant that the employer could no longer justify a deduction for liquidated damages. The employer's appeal was allowed. The judge stated that: 'If the conditions for the deduction of LADs [liquidated and ascertained damages] from a payment certificate are satisfied at the time when the Employer gives notice of intention to deduct, then the Employer is entitled to deduct the amount of LADs specified in the notice, even if the certificate of non-completion is cancelled by the subsequent grant of an extension of time.' The employer must, however, repay the additional amount deducted within a reasonable time.

4.65   In *Department of Environment for Northern Ireland* v *Farrans*, it was decided that the contractor has the right to interest on any repaid liquidated damages. This decision, however, was on JCT63; given that the SBC16 clauses expressly refer to repayment without stipulating that interest is due, it would appear that interest would not now be due in this situation. This was the view taken by His Honour Judge Carr in the first instance decision of *Finnegan* v *Community Housing Association* (1993) 65 BLR 103 (at page 114).

> *Department of Environment for Northern Ireland* v *Farrans (Construction) Ltd* (1982) 19 BLR 1 (NI)
>
> Farrans was employed to build an office block under JCT63. The original date for completion was 24 May 1975, but this was subsequently extended to 3 November 1977. During the course of the contract the architect issued four certificates of non-completion. By 18 July 1977 the employer had deducted £197,000 in liquidated damages but, following the second non-completion certificate, repaid £77,900 of those deductions. This process was repeated following the issue of the subsequent non-completion certificates. Farrans brought proceedings in the High Court of Justice in Northern Ireland, claiming interest on the sums that had been subsequently repaid. The court found for the contractor, stating that the employer had been in breach of contract in deducting monies on the basis of the first, second and third certificates, and that the contractor was entitled to interest as a result. The BLR commentary should be noted, which questions whether a deduction of liquidated damages empowered by clause 24.2 can be considered a breach of contract retrospectively. However, the case has not been overruled.

4.66   Certificates should always show the full amount due to the contractor. It is the employer alone that makes the deduction of liquidated damages. The employer would not be considered to have waived its claim by a failure to deduct damages from the first or any certificate under which this could validly be done, and would always be able to reclaim them as a debt at any point up until the final certificate.

5 Control of the works

5.1 The contract administrator's duties to the employer are normally set out in an appointment document, frequently on a standard form produced by one of the professional institutions. The contract administrator's role within the construction contract, however, is to be determined solely from the construction contract terms. A contract administrator named in SBC16 will be required to take certain actions; for example, supplying necessary information, issuing instructions and issuing certificates or statements. In some matters the contract administrator will act as agent of the employer, such as when issuing instructions which vary the works, and in others as an independent decision maker, for example when issuing certificates or deciding on claims for direct loss and/or expense. Failure by the contract administrator to comply with any obligation (usually prefaced by the phrase 'the contract administrator shall') will constitute failure on the part of the employer tantamount to breach of contract (see Tables 5.1 and 5.2). Furthermore, clause 1.11 now requires that all consents or approvals shall not be unreasonably delayed or withheld (except in relation to accepting defective work, and clause 7.1, assignment).

Table 5.1 Key powers of the contract administrator

Clause	Powers of the contract administrator
2.20.2	Request design documents from the contractor by a particular date
2.24	Consent to removal of goods, etc. from site
2.28.4	Fix an earlier completion date
2.28.5	Following practical completion, fix an earlier or later completion date
2.38.2	Issue instructions requiring defects to be made good
3.7.1	Consent to sub-contracting
3.8.2	Add persons to the list of sub-contractors
3.11	Issue notice requiring compliance with an instruction
3.12.2	Confirm an oral instruction in writing
3.14.1	Issue instructions requiring a variation
3.14.4	Sanction in writing any variation made by the contractor
3.15	Issue instructions postponing work
3.17	Issue instructions requiring inspections or tests
3.18.1	Issue instructions requiring removal of work from site
3.18.2	Allow non-compliant work, etc. to remain
3.18.3	Issue instructions requiring variation resulting from non-compliant work

Table 5.1 Key powers of the contract administrator – Continued

Clause	Powers of the contract administrator
3.18.4	Issue instructions requiring further inspections or tests
3.19	Issue instructions in relation to manner of carrying out work, including compliance with construction phase plan
3.21	Issue instructions excluding employed persons from site
4.12.1.1	Issue a pay less notice on behalf of the employer
4.25.3	Give the contractor one month's notice requiring supply of information reasonably necessary for assessing final account
5.3.1	Require a Schedule 2 quotation
6.5.1	Instruct clause 6.5.1 insurance is taken out
8.4.1	Give notice to contractor specifying defaults
Schedule 8	
9.6	Give instructions following termination of named specialist's employment

Table 5.2 Key duties of the contract administrator

Clause	Duties of the contract administrator
1.7	Communicate notices, etc. by agreed means
1.8	Issue certificates to the employer and the contractor at the same time
1.11	Ensure consents, approvals, etc. not unreasonably delayed or withheld
2.33	Issue notice regarding partial possession
2.8.2	Provide contractor with copies of contract documents, unless communications protocol requires otherwise
2.9.1.1	Provide contractor with copies of schedules and pre-construction information required by CDM Regulations
2.10	Determine levels and provide contractor with setting-out drawings
2.11	Provide contractor with copies of information on information release schedule
2.12.1	Provide contractor with copies of 'reasonably necessary' information
2.15	Issue instruction in relation to discrepancies
2.16.2	Notify contractor of agreement or decision regarding CDP discrepancy
2.17.1	Notify the contractor of any discrepancy between documents and statutory requirements
2.17.2	Issue instructions relating to clause 2.17.1 discrepancy
2.23.2	Notify contractor of potential infringement of patent rights
2.28.1	Give extensions of time
2.28.2	Notify contractor of decision regarding extensions of time
2.28.3	Apportion extension of time between relevant events

Table 5.2 Key duties of the contract administrator – Continued

Clause	Duties of the contract administrator
2.28.5	Fix a new completion date or confirm date already fixed
2.30	Issue a practical completion certificate for the works or a section
2.31	Issue a non-completion certificate for the works or a section
2.35	Issue a certificate confirming that defects in the relevant part have been made good
2.38.1	Issue a schedule of defects that appear during the rectification period
2.39	Issue a certificate of making good
3.10	Notify contractor on request of provisions empowering an instruction
3.16	Issue instructions in relation to provisional sums
3.18.2	Confirm decision to allow non-compliant work, etc. to remain
3.20	Give reasons for dissatisfaction with work within a reasonable time
3.22.2	Issue instructions regarding antiquities
4.9.1	Issue interim certificates
4.17.2	Prepare a statement specifying amount of retention deducted
4.18.1	Prepare a statement specifying amount of retention that would have been deducted
4.21.4	Notify the contractor of the ascertained amount of loss and/or expense
4.25.2.1	Ascertain amount of loss and/or expense due
4.25.2	Send copy of statement of final adjustment prepared by quantity surveyor to contractor
4.26.1	Issue final certificate
8.7.4	Issue a certificate setting out an account of balance due (unless employer issues a statement)
Schedule 1	
2	Return a copy of each design document submitted marked A, B or C
4	Identify why the contract administrator considers design document is not in accordance with the contract
7	Confirm or withdraw comment following notification by contractor
Schedule 2	
1.1	Supply contractor with further information to prepare quotation
4	Notify the contractor whether or not the variation or acceleration quotation is accepted
Schedule 4	Endeavour to agree the amount and method of opening up and testing; consider the stated criteria when issuing instructions
Schedule 8	
3.3	Issue instruction confirming agreed change
9.3	Issue instructions where contractor is unable to enter into a contract with a named specialist
9.4	Issue further instructions in relation to named specialists

5.2   Direct control over the carrying out of the contract works, including the manner in which the works are undertaken, is, however, solely the responsibility of the main contractor (cl 3.6). The duty of the contract administrator to the employer will normally be to inspect the work at intervals. The contractor is obliged to provide the contract administrator with reasonable access to the works, its premises, and to its sub-contractors' premises (cl 3.1). The precise obligation and purpose of such visits will arise directly from the terms (either express or implied) of the professional appointment, and of course SBC16 includes no express provision relating to inspection or monitoring of work by the contract administrator. Clearly, when the contract administrator is required under the contract to form an opinion on various matters, including the standard of work and materials prior to issuing a certificate, then it would be implied, even if not expressly set out in the terms of appointment, that some form of inspection must take place.

Site manager and contractor's persons

5.3   The contractor is required to keep a competent 'Site Manager' on the site 'at all material times' (cl 3.2). There is no requirement in the contract conditions to have the person named in the contract documents; however, they must be approved by the employer. This is an important role, as this person may receive instructions from the contract administrator, and therefore acts as the contractor's agent. It would therefore be good practice to establish the identity of the site manager no later than at the pre-contract meeting, and to make sure this is agreed and recorded in writing. If the contractor appoints a replacement, this must also be approved by the employer (cl 3.2). The contractor has full responsibility for the performance of the site manager and all people the site manager engages on the project (cl 3.6).

5.4   The contract administrator has the power to exclude persons from site (cl 3.21), but may not do so 'unreasonably or vexatiously'. This power is only likely to be used where it appears that the employee may be seriously affecting operations on site. If used unreasonably, this could constitute an 'impediment' under clauses 2.29.7 and 4.22.5.

Employer's representative

5.5   Although the majority of the administration of the contract is dealt with by the contract administrator, the employer has an active role to play. The employer is required to make decisions on various matters and to issue notices direct to the contractor and is entitled to exercise various powers (see Tables 5.3 and 5.4). The employer is entitled to appoint a representative to exercise all these functions (cl 3.3). The contractor must be notified in

Table 5.3 Key duties of the employer	
Clause	Duties of the employer
1.7	Communicate notices, etc. by agreed means
1.11	Ensure consents, approvals, etc. not unreasonably delayed or withheld
2.4	Give possession of the site
3.5.1	Nominate a replacement contract administrator

Table 5.3 Key duties of the employer – Continued

Clause	Duties of the employer
3.23	Comply with CDM Regulations
3.23.1	Ensure principal designer and principal contractor (if not the contractor) comply with CDM Regulations
3.23.4	Notify the contractor if the principal designer or principal contractor is replaced
4.5.1	Pay VAT properly chargeable
4.11.2	Pay contractor amount stated in the certificate
4.11.3	Pay contractor the amount stated in the payment notice
4.11.5	Issue pay less notice to contractor if intending to pay less than the certified sum
4.11.6	Pay interest to the contractor on unpaid amounts
4.17.3	Place retention in a separate banking account and certify that this has been done
6.9.1	Ensure works insurance policy provides for recognition of, or waives any right of subrogation against, any sub-contractor
6.10.1	Take out terrorism cover
6.11.1	Notify the contractor if notified that terrorism cover has ceased
6.11.2	Notify the contractor whether employment is to continue or terminate
6.12.1	Provide evidence of insurance to the contractor
6.13.5.1	Pay all monies from insurance to contractor
6.17	Comply with the Joint Fire Code, and ensure all employer's persons comply
8.7.4	Issue a statement setting out an account of balance due (unless the contract administrator issues a certificate)
8.7.5	Pay the contractor any balance due
8.8.1	Notify the contractor of decision not to complete the works; send the contractor a statement of the balance due
8.10.2	Give the contractor notice if it makes any proposal, etc. in relation to insolvency
8.12.3	Prepare an account (if contractor not required to do so)
8.12.5	Pay the contractor the amount properly due
Schedule 3	
B.1	Take out and maintain a joint names policy for full reinstatement value of the works
C.1	Take out and maintain a joint names policy in respect of existing structures
C.2	Take out and maintain a joint names policy for full reinstatement value of the works
Schedule 8	
5	Monitor contractor's performance by reference to indicators
6	Promptly notify the contractor of matters likely to give rise to a dispute, and meet as soon as possible for good faith negotiations

Table 5.4	Key powers of the employer
Clause	**Powers of the employer**
2.5	Defer possession of the site
2.6.1	Use or occupy the site
2.7	Have work executed by employer's persons
2.32.1	Notify the contractor of intention to deduct liquidated damages
2.33	Take possession of part of the works prior to practical completion, with contractor's consent
3.3	Appoint a representative, terminate such appointment and appoint a replacement
3.4	Appoint a clerk of works
3.7.1.2	Consent to sub-contracting contractor's designed portion
3.8.1	Add persons to the list of sub-contractors
3.11	Employ and pay others to carry out work
6.11.2	Terminate the contractor's employment
6.12.2	Take out insurance if contractor defaults and deduct amount from monies due
6.13.5.2	Retain amounts for professional fees from insurance monies payable to contractor
6.14	Terminate the contractor's employment
7.2	Assign the right to bring proceedings to any transferee
7A.1	Give notice stating that third party rights shall vest in a purchaser or tenant
7B.1	Give notice stating that third party rights shall vest in a funder
7C	Give notice requiring contractor to enter into a warranty with a purchaser or tenant
7D	Give notice requiring contractor to enter into a warranty with a funder
7E.1	Give notice requiring contractor to comply with requirements set out in contract particulars as to obtaining sub-contractor warranties with purchasers, tenants/funders or the employer
8.4.2	Terminate the contractor's employment because of continuation of specified default
8.4.3	Terminate the contractor's employment because of repeat of specified default
8.5.1	Terminate the contractor's employment because of contractor insolvency
8.5.3.3	Take reasonable measures to ensure that the site, etc. is protected
8.6	Terminate the contractor's employment because of corruption or where regulation 73(b)(1) of the PC Regulations applies
8.7.1	Employ and pay other persons to carry out and complete the works; enter upon the site and use temporary buildings, etc.
8.11.1	Terminate the contractor's employment because of suspension of the works
9.4.1	Give notice of arbitration
9.4.3	Give further notice of arbitration
9.7	Apply to the courts to determine a question of law
Schedule 8	
5.3	Inform the contractor that performance targets may not be met
7.1	Publish amendments to the JCT standard form contract

writing of the identity of the individual, and of any exceptions to the functions the individual will perform. Footnote [39] to the form makes it clear that to avoid any confusion in the roles, neither the contract administrator nor the quantity surveyor should be appointed as the employer's representative.

Clerk of works

5.6   The employer is entitled to employ an independent clerk of works whose duty is solely 'to act as inspector on behalf of the Employer under the Architect/Contract Administrator's directions' (cl 3.4).

5.7   It should be noted that the presence of a clerk of works does not lessen the contract administrator's duty in respect of site inspection (*Kensington Health Authority* v *Wettern*). The clerk of works does not act as agent for the contract administrator but may issue directions to the contractor. However, any such direction must be one which the contract administrator could have made under the contract and must be confirmed in writing by the contract administrator within two working days if it is to be effective. The contractor will frequently take the initiative and ask for an instruction to confirm the clerk of works' direction, but there is no requirement for the contractor to do this, and therefore the contract administrator and clerk of works should be careful to co-ordinate over such matters. To avoid confusion, it may be sensible for the clerk of works to refrain from giving an instruction directly, unless the matter is urgent, but instead to request that the contract administrator issues it; alternatively, all clerk of works' instructions could be given in writing, with copies sent at the same time to the contract administrator.

> **Kensington and Chelsea and Westminster Area Health Authority v Wettern Composites**
> **(1984) 31 BLR 57**
>
> Wettern Composites was the subcontractor for the supply and erection of precast concrete mullions for an extension to the Westminster Hospital, on which the Health Authority had also engaged architects, engineers and a clerk of works. Tersons Ltd was the main contractor. The hospital was completed in 1965. In 1976 it was discovered that there were considerable defects in the mullions. The Health Authority brought an action against the architects, engineers and sub-contractor, though the latter subsequently went into liquidation. Judgment was given for the Health Authority. The architects had failed to exercise reasonable skill and care in ensuring conformity of the works to the design. Although a clerk of works had been employed, this did not lessen the architects' responsibility. However, the Health Authority was vicariously liable for the contributory negligence of its clerk of works, and the damages recoverable from the architects were reduced by 20 per cent accordingly.

Principal contractor

5.8   The contract assumes that the contractor will act as principal contractor for the purposes of the CDM Regulations, unless another firm is named in Article 6. It is the employer's responsibility to appoint a principal contractor; therefore, if the contractor is unable to or ceases to take on this role, the employer must appoint a substitute (Article 6). It is the contractor's responsibility to develop the construction phase plan so that it complies with the Regulations (regulation 12), and to ensure that the works are carried out in accordance with the plan.

5.9   The principal designer has no duty to inspect the works and would be very unlikely to visit the site unless some highly unusual circumstance arises, such as the discovery of an unanticipated hazard. The main responsibility for ensuring that correct health and safety measures are employed on site rests with the contractor, both under statute and under the express terms of the contract.

Information to be provided by the contract administrator

5.10   In the majority of construction contracts the information contained in the contract documents will not be sufficient to enable the project to be constructed. Even if the works have been fully specified it is likely, for example, that information regarding assembly, location, detail dimensions, colours, etc. will be needed by the contractor throughout the project. Supply of this information will usually form part of the contract administrator's duties to the employer under the terms of appointment.

5.11   SBC16 refers in four places to the contract administrator's obligation to provide information. These refer to 'descriptive schedules or similar documents' and information required under the CDM Regulations (cl 2.9.1.1); setting-out information (cl 2.10); 'information referred to in the Information Release Schedule' (cl 2.11); and 'such further drawings or details as are reasonably necessary to explain and amplify the Contract Drawings' (cl 2.12.1). Although the clauses do not require this information to be released under a contract administrator's instruction, this is common practice – and is wise, as it would enable the clause 3.11 provisions to be brought into operation if necessary (see paragraph 5.32 on contract administrator's instructions). If any of the information supplied introduces changes or additions to the works, it must be covered by a contract administrator's instruction requiring a variation.

5.12   The 'descriptive schedules or similar documents' and CDM regulation 4 information are to be provided as soon as possible after the execution of the contract (cl 2.9.1.1). These appear to be by way of amplification of information given in the tender documentation, as the clause makes it clear that nothing in these documents shall 'impose any obligation beyond those imposed by the Contract Documents' (cl 2.9.3).

5.13   Under clause 2.10 the contract administrator is responsible for giving sufficient accurately dimensioned drawings and levels to enable the contractor to set out the works, and the contractor must accurately follow this information. The contractor must 'amend any errors' that result from its own inaccurate setting out. Alternatively, the contract administrator, with the employer's consent, may instruct that the error remains, in which case 'an appropriate deduction may be made from the Contract Sum' (cl 2.10). There is no suggestion in the conditions as to how this might be assessed, and in practice it will be a matter for negotiation. The error and the deduction should first be discussed with the employer, and the agreed deduction should ensure adequate compensation.

5.14   Information shown on the information release schedule must be supplied at the stipulated date, unless the contract administrator is prevented from doing so 'by an act or default of the Contractor' (cl 2.11). Failure to provide the information and instructions under clause 2.11 would constitute a 'Relevant Event' under clause 2.29.7 and a 'Relevant Matter' which may give rise to a direct loss and/or expense claim under clause 4.22.5, and possible grounds for termination under clause 8.9, but only where such failure has led to the suspension of the carrying out of the whole of the works for at least a period which has been entered in the contract particulars (cl 8.9.2). An act or default of the contractor might include, for example, failure to provide design documents as required by the contract and which the contract administrator needs to finalise part of the design.

5.15 There is no mechanism whereby the contract administrator may unilaterally adjust the schedule following, for example, an extension of time. Such adjustments will have to be negotiated and agreed by the parties, and it may be necessary to do this on a regular basis, keeping the contract administrator involved (the contract allows the employer and contractor to agree changes to the information release schedule under clause 2.11). Parties should tackle this co-operatively – the contract states that such agreement should not be unreasonably withheld. For example, if variations have been issued that involve additional work and have resulted in an extension of time, or if work has been omitted and an earlier completion date fixed, then it would be reasonable for the information release schedule to be adjusted to reflect this. If the contractor refuses to agree to an adjustment, the document will become worthless with respect to assessing extensions of time.

5.16 With respect to information not shown on the information release schedule, or where a schedule is not used, the contract administrator is under an obligation to provide 'such further drawings or details as are reasonably necessary to explain and amplify the Contract Drawings and shall issue such instructions … as are necessary to enable the Contractor to carry out and complete the Works' (cl 2.12.1). It is suggested that this obligation would extend to both amplification of information in the contract documents and providing full information regarding any variation that is required to be carried out. The inclusion of the word 'reasonably' suggests that the contractor can be expected to obtain some detailed information, for example manufacturers' fixing information. The contract administrator should be careful, however, in respect of leaving decisions to the contractor, as it may not always be possible to hold the contractor responsible should a detail or fixing fail.

5.17 The information and instructions should 'be provided at the time the Contractor reasonably requires them, having regard to the progress of the Works'. However, there is no need to supply them any earlier than would be necessary to allow the contractor to complete by the completion date (cl 2.12.2). There is no general requirement for the contractor to apply for information in writing, but if the contractor has 'reason to believe that the Architect/ Contract Administrator is not aware' of when information may be needed, the contractor should give 'advance notice' to the contract administrator (cl 2.12.3). In practice, such notice may frequently be in the form of a programme indicating dates when information is required. As above, failure to provide the information and instructions under clause 2.12.1 would constitute a relevant event under clause 2.29.7 and a relevant matter which may give rise to a direct loss and/or expense claim under clause 4.22.5, and possible grounds for termination under clause 8.9.2.

5.18 Whether or not an information release schedule is used, if any acceleration to the works is proposed (e.g. under the acceptance of a Schedule 2 quotation), the contract administrator should take care to warn the employer at an early stage if this might present difficulties for programming of information.

5.19 Under SBC16 there is no provision for other consultants to issue information direct to the contractor; this would have to be done through the contract administrator. Delay in supplying necessary drawings by other consultants would therefore have the effect under the contract of a delay on the part of the contract administrator, i.e. a delay for which the employer is responsible. An obligation to supply information on time would normally be implied into the terms of engagement of any consultant, if not expressly set out (*Royal Brompton Hospital* v *Frederick Alexander Hammond*).

Royal Brompton Hospital National Health Trust v *Frederick Alexander Hammond and others* (No. 4) [2000] BLR 75

The Royal Brompton Hospital (RBH) engaged Frederick Alexander Hammond to undertake a £19 million construction project on a JCT80 standard form of contract. The contractor successfully claimed against RBH, including for losses suffered due to delays. RBH commenced proceedings against 16 defendants, who were all members of the professional team. A trial was fixed to deal with a number of different issues, all of which were settled except for one relating to the consulting M&E engineers, Austen Associates Ltd (AA). The issue was whether AA was obliged to provide co-ordination and builder's work information to ensure that RBH complied with clause 5.4 of the main contract. The court decided that AA was under a duty to use reasonable skill and care to ensure that the drawings were provided in time to enable the contractor to prepare his installation drawings, and thus to carry out and complete the works in accordance with the contract conditions.

Information provided by the contractor

5.20   The contractor may be required to provide information in regard to requirements of the CDM Regulations, and in relation to the completion of design for which the contractor is responsible under the contractor's designed portion of the works.

5.21   The contractor as 'Principal Contractor' may be required by the principal designer to provide information in relation to the health and safety file (regulation 7(12) and clause 3.23.2). If the contractor is acting as 'Principal Designer' then, under regulation 12(5) and (6), it is required to prepare and deliver the health and safety file to the employer.

5.22   Under clause 2.40, the contractor is obliged to provide the employer before practical completion with 'such Contractor's Design Documents and related information as is specified in the Contract Documents or as the Employer may reasonably require'. The documents and information are those that 'show or describe the Contractor's Designed Portion as built or relate to the maintenance and operation of it or its installations'. Note that if nothing is set out in the contract documents the contractor is still under an obligation to provide such information as may reasonably be required which, depending on the scale of the designed portion, could amount to a substantial operation and maintenance manual.

Contractor's design submissions

5.23   SBC16 contains detailed provision regarding the submission of the developing design by the contractor. This information is essential in order for the contract administrator and employer to monitor the development of the design and to integrate it with the rest of the works.

5.24   The contractor must provide the contract administrator with 'such Contractor's Design Documents as are reasonably necessary to explain or amplify the Contractor's Proposals' (cl 2.9.4). This information should include (if requested) related calculations and information (cl 2.9.4.1) and (whether or not requested) 'all levels and setting out dimensions which the Contractor prepares or uses for the purposes of carrying out and completing the Contractor's Designed Portion' (cl 2.9.4.2). 'Contractor's Design Documents' are defined as 'the drawings, details and specifications of materials, goods and workmanship and other related documents and information prepared by or for the Contractor in relation to the Contractor's Designed Portion (including such as are contained in the Contractor's Proposals or referred to in clause 2.9.4), together, where applicable, with any other design documents or information to be provided by him under the BIM Protocol' (cl 1.1). In practice there could be differences of opinion as to what information may be 'reasonably necessary'. The information the contractor may need to actually construct the work may be different from the information that the employer and contract administrator would like to receive. If specific information is needed, and the BIM protocol is not used, then it may be sensible to set out a schedule in the employer's requirements.

5.25   As regards timing, the information is to be provided 'in accordance with the Design Submission Procedure', and 'the Contractor shall not commence any work to which such a document relates before that procedure has been complied with' (cl 2.9.5). The 'Design Submission Procedure' is defined in clause 1.1 as 'such procedure as is specified in the BIM Protocol or, where that is not applicable, the procedure set out in Schedule 1, subject to any modifications of that procedure set out in the Contract Documents'.

Schedule 1 procedure

5.26   The design submission procedure (Schedule 1) states that the documents should be submitted 'by the means and in the format stated in the Employer's Requirements'. It also states that documents must be submitted 'in sufficient time to allow any comments of the Architect/Contract Administrator to be incorporated' (Schedule 1:1) and refers to 'the period for submission … stated in the Contract Documents' (Schedule 1:2). It would therefore be open to the employer – and on most projects it would be wise – to include detailed requirements regarding scope, format and timing of submissions in the contract documents.

5.27   Following submission of a contractor's design document, the contract administrator must respond within 14 days of the date of receipt, 'or (if later) 14 days from either the date or expiry of the period for submission of the same stated in the Contract Documents' (Schedule 1:2) (in other words, if the contractor supplies information earlier than any agreed date, the contract administrator would not be required to respond any earlier than stated in the contract documents).

5.28   The contract administrator is entitled to take three alternative courses of action (see Figure 5.1); it can accept the contractor's design document, in which case it should return it marked 'A'. It may accept it, subject to certain comments being incorporated, in which case it should be marked 'B'. Or it can make comments and require the contractor to resubmit the document with the comments incorporated for further approval, in which case it should be marked 'C' (Schedule 1:5). In the case of marking B or C, the contract administrator must state why the document does not comply with the contract (comments are only valid if the document does not comply (Schedule 1:2) – if it does comply, any required alteration would constitute a variation). If the contract administrator does not respond within the specified period, it is deemed to have accepted the document (Schedule 1:3).

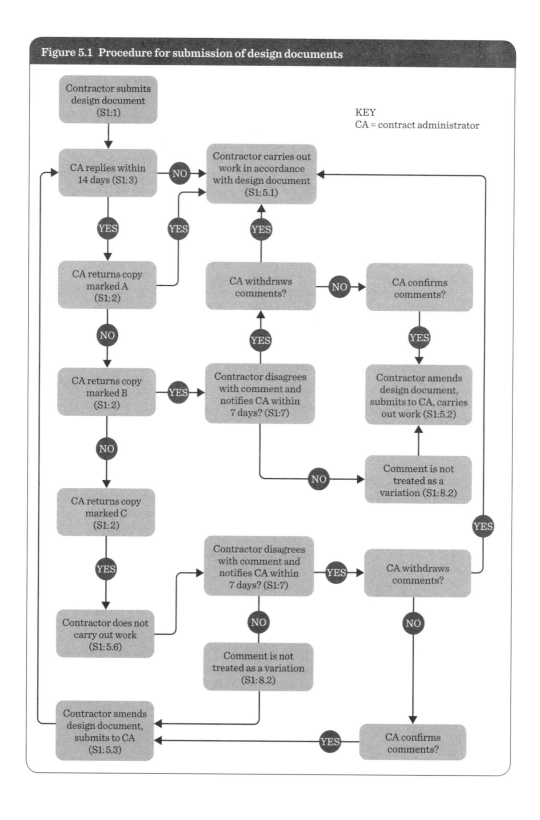

Figure 5.1 Procedure for submission of design documents

5.29 Schedule 1:8.3 states that no comments or any action by the contract administrator will relieve the contractor of its liability to ensure that the document complies with the contract, or that the project complies with the contract. This has the effect that if the contractor incorporates a comment made by the contract administrator then it accepts that the comment has been properly made (i.e. it identifies a way in which the design document is not in accordance with the contract).

5.30 If the contractor disagrees with a comment and considers that the document complies with the contract, it is required to inform the contract administrator, within seven days of receipt of the comment, that compliance with the comment would give rise to a variation (Schedule 1:7). The contractor must give reasons as to why it thinks this. The contract administrator must either confirm or withdraw the comment within seven days. The confirmation or withdrawal is stated not to signify that the employer accepts that the contractor's design document complies or that the comment represents a variation (Schedule 1:8.1) – this would be a question of fact, if necessary to be resolved by adjudication. The contractor would have to implement the comment and argue its case later.

5.31 If the contractor does not notify the contract administrator of its disagreement with a comment, then the comment will not be treated as a variation, even if it could be later shown in fact to be a variation (Schedule 1:8.2).

Contract administrator's instructions

5.32 Only the contract administrator is given the power to issue instructions (cl 3.10). Sometimes the contract administrator 'may' issue instructions (e.g. instructions requiring a variation under clause 3.14), but at other times the contract administrator 'shall' issue instructions (e.g. instructions regarding discrepancies between contract documents under clause 2.15). The latter is an obligation. If the employer gives an instruction other than through the contract administrator, this would not be effective under the contract. The contractor would be under no obligation to comply with any such instruction. If the contractor does, however, carry out the instruction, a court might consider that there had been an agreed amendment to the contract. The consequences of such an agreement would be difficult to sort out in practice – the employer would be very unwise to make such agreements or issue any instructions other than through the contract administrator.

5.33 Clause 3.12.1 states that if any instruction is given 'otherwise than in writing' (which would normally mean oral instructions), it is of no immediate effect, but the contractor must confirm its terms in writing within seven days. It then takes effect after seven days have elapsed from the date of the contractor's confirmation, provided that the contract administrator has not, in the meantime, dissented in writing. The contract administrator must therefore check carefully that the confirmation represents exactly what is required. Of course, the contract administrator may choose to confirm the instruction, in which case it takes effect from the date of the contract administrator's confirmation (cl 3.12.2). If neither party confirms the oral instruction, but the contractor acts upon it anyway, then the contract administrator can later sanction the instruction at any time prior to the issue of the final certificate (cl 3.12.3), but the contractor would be taking a risk (*MOD* v *Scott Wilson Kirkpatrick and Partners*).

> *Ministry of Defence v Scott Wilson Kirkpatrick and Partners* [2000] BLR 20 (CA)
>
> Scott Wilson Kirkpatrick (SWK) was engaged as structural engineer and supervising officer by the MOD in relation to refurbishment of the roof at Plymouth Dockyard under GC/Works/1. Several years after the works were completed, wind lifted a large section of roof and deposited it in a nearby playing field. The contract required 9–12 in. nails, but the contractor had used 4 in. nails. The supervising officer had been party to discussions regarding the use of the 4 in. nails, but neither he nor the contractor could remember very clearly when these discussions had taken place, or exactly what had been said. The Court of Appeal decided that the evidence was sparse and vague, and declined to find that there was any instruction under 7(1)(a) or 7(1)(m) (instructions that may be given orally), or that there had been any agreement as to the replacement. Even if the supervising officer's conduct amounted to confirmation or encouragement, this could not absolve the contractor from its duty to fix the purlins in a workmanlike manner. The MOD was therefore entitled to insist on its strict contractual rights. The Court of Appeal noted, however, that an instruction in writing was not a condition precedent to a claim by the contractor, so long as it was able to prove that the change had been agreed.

5.34   All instructions must be in writing and sent in the format and by the means the parties 'have agreed or may from time to time agree in writing', which could include electronic communications (cl 1.7.2). If no means have been agreed, instructions may be sent by 'any effective means' (cl 1.7.3), and will be considered 'duly served' if sent by the methods set out in clauses 1.7.3.1 and 1.7.3.2. It should be remembered that it may also be necessary to prove when an instruction has been received, and therefore it is advisable to send a hard copy by recorded delivery or to record receipt of instructions at a subsequent progress meeting.

5.35   Unless a special format has been agreed, an instruction in a letter would be effective, as long as the letter is quite clear. A drawing sent with a letter requiring it to be executed would constitute an instruction, but a drawing with no covering instruction may be judged to be ineffective. Instructions in site meeting minutes may also constitute a written confirmation of an oral instruction if issued by the contract administrator. It would depend on the circumstances whether the minutes were sufficiently clear to fall within the terms of the contract, and it is therefore not good practice to rely on this method. The use of site instruction books should also be avoided. Although signing an instruction in a book would constitute a written instruction under the terms of the contract, there is no obligation to sign such books, and it may also be prudent not to make quick decisions on site but to wait until all the implications of the instruction can be considered. In practice, many people use the instruction forms published by the RIBA (either as hard copy or through NBS Contract Administrator).

5.36   The contractor must comply with every instruction (see Table 5.5) provided that it is valid, i.e. provided that it is in respect of a matter regarding which the contract administrator is empowered to issue instructions. The contractor must 'forthwith' comply, which for practical purposes means as soon as is reasonably possible (cl 3.10).

5.37   If the contractor feels that a contract administrator's instruction might not be empowered by the contract, or requires clarification, then the contractor may ask the contract administrator to specify in writing the provisions of the contract under which the instruction is given, and the contract administrator must do this 'forthwith' (cl 3.13). The contractor must then either comply or issue a notice disputing the matter. If, however, the contractor chooses to accept the contract administrator's reply and complies with the instruction,

Control of the works 79

Table 5.5 Matters about which the contract administrator is empowered to issue instructions

Contract administrator may/shall issue instructions:	Clause	Power/duty
that errors in setting out shall not be corrected	2.10	power
necessary to enable contractor to carry out works	2.12.1	duty
in regard to clause 2.14 discrepancies	2.15	duty
in regard to clause 2.14 discrepancies	2.16.1	duty
in regard to clause 2.17.1 discrepancies (statutory requirements)	2.17.2	duty
requiring defects to be made good	2.38.2	power
confirming clerk of works directions	3.4	power
varying the works, etc.	3.14.1	power
postponing work	3.15	power
relating to provisional sums	3.16	duty
requiring inspections or tests	3.17	power
requiring removal of work from site	3.18.1	power
requiring a variation resulting from non-compliant work	3.18.3	power
requiring further inspections or tests	3.18.4	power
relating to the manner of carrying out work, including compliance with construction phase plan	3.19	power
excluding employed persons from site	3.21	power
relating to antiquities	3.22	duty
requiring remedial measures	6.19.1.2	duty
Schedule 2		
confirming that a variation quotation or acceleration quotation is accepted	4	power
confirming whether or not a variation is to be carried out	5	duty
Schedule 8		
where contractor is unable to enter into a contract with a named specialist	9.3	duty
in relation to named specialists	9.4	duty

then the employer is bound by the instruction. This would appear to be the case even if at a later stage it is established that the contract administrator had no authority under the contract.

5.38 Even if the contractor decides to query the instruction under clause 3.13, this does not relieve the contractor of the obligation to comply. Should the instruction be found to be valid, the contractor would be liable for any delay caused by failing to comply as required by the contract. If the contractor does comply, but the instruction turns out to have been invalid, the contractor may be entitled to any losses incurred through compliance.

5.39   The requirement to comply with a valid instruction is subject to certain exceptions:

- a clause 5.1.2 variation instruction (access and use of the site, etc.) to the extent that the contractor makes a reasonable objection (cl 3.10.1);

- an instruction relating to a Schedule 2 variation quotation, until a confirmed acceptance has been given (cl 3.10.2);

- where the instruction might affect the efficacy of the design of the Contractor's Designed Portion (cl 3.10.3, see paragraph 5.46);

- where the instruction might affect the contractor's compliance with the CDM Regulations (cl 3.10.3, see paragraph 5.40);

- where the instruction may infringe patent rights (cl 3.10.4);

- where the instruction relates to a named specialist, and the contractor is unable to enter into a contract with that firm (cl 3.10.5, Schedule 8:9.3 and 9.4, see paragraph 5.69).

5.40   In respect of the exceptions under clause 3.10.3, and provided the contractor notifies the contract administrator of the problem within seven days, the instruction will not take effect unless and until it is confirmed by the contract administrator (cl 3.10.3). In the case of patent right infringement, again the contractor must notify the contract administrator (although there is no time limit for this), and vice versa if the contract administrator is the party who becomes aware of the infringement, and the instruction will not take effect until confirmed (cl 2.23.2).

5.41   If the contractor does not comply with a written instruction, the employer may employ and pay others to carry out the work to the extent necessary to give effect to the instruction (cl 3.11). The contract administrator must have given written notice to the contractor requiring compliance with the instruction, and seven days must have elapsed after the contractor's receipt of the notice before the employer may bring in others. This indicates that a recorded form of delivery is desirable. Although there is no obligation to issue such notices, it would be prudent for the employer to take swift action in order to protect its interests. The employer is entitled to recover any additional costs from the contractor, i.e. the difference between what would have been paid to the contractor for the instructed work and the costs actually incurred by the employer. These costs could include not only the carrying out of the instructed work, but also any special site provisions that need to be made, including health and safety provisions, and any additional professional fees incurred. Although it would be wise to obtain alternative estimates for all these costs wherever possible, if the work is urgent there would be no need to do so.

Variations

5.42   Contract administrator's instructions often require some variation to the works. Under common law neither party to a contract has the power to unilaterally alter any of its terms. Therefore, in a construction contract neither the employer nor the contract administrator would have the power to require any variations unless the contract contains such a power. As some aspects of construction may be difficult to define exactly in advance, most construction contracts contain provisions allowing the employer to vary the works to some degree. Changes can arise because of a variety of reasons, including unexpected site problems or because of design changes wanted by the employer or which become necessary in order to integrate the contractor's designed portion.

5.43   Under SBC16 the contract administrator is empowered to order specific variations (cl 3.14.1). The scope of what constitutes a variation is set out in clause 5.1. It is broadly defined and includes alterations to the quantity and specification of the works, and to operational restrictions such as access to the site. The contract expressly states that no variation will vitiate the contract (cl 3.14.5), but the power does not extend to altering the nature of the contract, nor can the contract administrator issue variations after practical completion. All variations under clause 3.14.1 may result in an adjustment of the contract sum (cl 4.3.1) and give rise to a claim for an extension of time (cl 2.29.1) or direct loss and/or expense (cl 4.22.1). If the works are suspended as a consequence, the variation may also be grounds for termination by the contractor, unless the variation is necessitated by some negligence or default of the contractor (cl 8.9.2).

5.44   The contract administrator may vary the works (cl 5.1.1), for example by changing the standard of a material specified. The contract administrator may add to or omit work, substitute one type of work for another or remove work already carried out. The contract administrator may vary the access to or use of the site, limitations on working space or working hours, the order in which the work is to be carried out or any restrictions already imposed (cl 5.1.2). As well as giving the employer a great deal of flexibility, this contract provision is necessary to accommodate difficulties that may arise, for example through local authority restrictions on working hours. However, the contractor need not comply with a clause 5.1.2 instruction to the extent that it makes reasonable objection (cl 3.14.2). Given that the contractor will be paid for such variations, it is difficult to see what might constitute a 'reasonable' objection, although it is possible that a variation might, for example, have a detrimental 'knock-on' effect on some other project, causing the contractor to suffer losses for which it would not otherwise be compensated.

5.45   Finally, the contract administrator may sanction any variation made by the contractor other than under an instruction of the contract administrator (cl 3.14.4). If such a variation were likely to affect the employer, the contract administrator would be wise to discuss it with the employer before taking action.

Variations to the contractor's designed portion

5.46   Clause 3.14.3 states that where instructions require a variation in respect of the contractor's designed portion, any instruction 'shall be an alteration to or modification of the Employer's Requirements'. This would appear to prevent the contract administrator from directly requiring changes to the proposals after the contract is entered into, including to any further design details that are developed as the contract progresses, except in cases where the developing design does not meet the employer's requirements. If the contract administrator issues any instruction which in the opinion of the contractor may affect the efficacy of the design, the contractor must object within seven days of receiving the instruction. The instruction will not then take effect until confirmed by the contract administrator (cl 3.10.3).

Defective work

5.47   Clauses 2.3.1 and 2.3.2 state that all materials, goods and workmanship shall be of the standard specified in the contract documents (see discussion under paragraphs 3.18–3.24). The contract administrator will normally inspect at regular intervals to monitor the standard that is being achieved. However, achieving the contractual standard is the responsibility of the contractor, and the lack of an inspection cannot be used as an excuse for sub-standard work. When the standard achieved appears to be unsatisfactory it can be tempting to become involved in directing the day-to-day activities of the contractor on site. Apart from being an enormous burden on the contract administrator, this could confuse the issue of who is ultimately responsible for quality and is to be avoided.

5.48   Where the standard of goods, materials or workmanship is a matter for the approval of the contract administrator under clause 2.3, any dissatisfaction should be expressed within a reasonable time (cl 3.20). The contract administrator must also state reasons for the dissatisfaction. In all other cases there is no obligation to point out defects or errors, but the contract administrator would normally, of course, draw the contractor's attention to areas of defective or poor quality work. The fact that work has been included in an interim certificate does not relieve the contractor of its responsibility for the standard of work, or prevent the contract administrator from deciding that the work is defective (cl 3.6).

Testing work

5.49   The contract administrator may instruct the contractor to open up completed work for inspection, or arrange for testing of any of the work or materials, fixed or unfixed (cl 3.17). No time limit is specified, but obviously the contract administrator should instruct as soon as the need for such action becomes apparent (delay could result in escalating or unnecessary costs). Failure to ask for tests in no way relieves the contractor of the obligation to provide work according to the contract. The cost of carrying out the tests is added to the contract sum, unless it was already provided for in the bills of quantities under a provisional sum or unless the work proves to be defective. Unless the work is defective, the contractor may also be entitled to an extension of time under clause 2.29.2.2 and loss and/or expense under clause 4.22.2.2, unless these tests were provided for in the contract bills.

5.50   The contract administrator has several courses of action if work is defective. An instruction can be issued requiring the removal of work, materials or goods from the site (cl 3.18.1); the work, materials or goods can be allowed to remain (cl 3.18.2) (unless part of the contractor's designed portion); a variation can be issued (cl 3.18.3); or further tests can be instructed 'having due regard' to the Code of Practice for tests set out in Schedule 4 (cl 3.18.4).

5.51   If a notice requiring compliance with a clause 3.18.1 instruction to remove from the site is given and not complied with, then the provisions of clause 3.11 could be brought into operation (*Bath and NE Somerset DC v Mowlem*). To fall under clause 3.18.1, the instruction must specifically require removal of the work from site, however impractical. Simply drawing attention to the defective work would not be sufficient (*Holland Hannen v Welsh Health Technical Services*). Refusal to remove defective work is also a ground for termination under clause 8.4.1.3, provided there has been a written notice or instruction and the refusal materially affects the work.

Bath and North East Somerset District Council v Mowlem plc [2004] BLR 153 (CA)

Mowlem plc was engaged on JCT98 (Local Authorities With Quantities) to undertake the Bath Spa project. Completion was expected to be in 2002 but work was still under way in 2003. Paint applied by Mowlem to the four pools began to peel, and the contract administrator issued architect's instruction no. 103 which required Mowlem to strip and repaint the affected areas. Mowlem refused to comply and the Council issued a notice under clause 4.1.2. Mowlem still did not comply, and the employer engaged Warings to carry out this work. Mowlem refused Warings access to the site, and the Council applied to the court for an injunction, which was granted. Mowlem appealed against the injunction, but the appeal was dismissed.

Mowlem had argued that it was able to rectify all the defects and that the liquidated damages provided under the contract were the agreed remedy for delays caused. The Council was able to show that the liquidated damages were not adequate compensation for the losses suffered. Lord Justice Mance held that, in such cases, the court should examine whether the liquidated damages would provide adequate compensation, and if they would not, as in this case, it was appropriate to grant an injunction. In reaching this decision he took into account irrecoverable losses such as the 'unquantifiable and uncompensatable damage to the Council's general public aims'.

Holland Hannen & Cubitts (Northern) Ltd v *Welsh Health Technical Services Organisation* (1985) 35 BLR 1 (CA)

Cubitts Ltd was employed by the Welsh Health Technical Services Organisation (WHTSO) to construct two hospitals at Rhyl and Gurnos. Percy Thomas (PTP) was the architect. Redpath Dorman Long Ltd (RDL) was the nominated sub-contractor for the design and supply of pre-cast concrete floor slabs. RDL assured WHTSO that the floors would be designed to CP 116 (concerning deflection), but the design team later required RDL to work to CP 204. Following installation, the contractor complained about extra work and costs due to adjustments to the partitions necessitated by excessive deflection of the floors, and it was established that they had been designed to CP 116 not CP 204. PTP sent three letters 'condemning' the floors, but the first did not mention clause 6(4), and none of them required removal of the work. Cubitts stopped work for 20 weeks until PTP issued instructions specifying how the defect should be resolved. Cubitts commenced proceedings, claiming compensation for delay. The claim was settled, but the relevant parties maintained their proceedings against each other for contribution. The official referee decided that RDL was liable for two-thirds of the amount paid to Cubitts and the design team for one-third. The Court of Appeal decided that this was incorrect and the correct apportionment should have been that RDL was liable for one-third and the design team for two-thirds. In reaching this conclusion it stated: 'PTP contributed very substantially to the delay which occurred, in failing to recognise the defect in the design at an earlier stage; by issuing an invalid notice in 1976, and by moving very slowly thereafter to take the necessary steps to have the defects in the flooring put right' (Robert Goff LJ).

5.52   If defective work is to be allowed to remain, there must be consultation with the contractor and the approval of the employer must be obtained (cl 3.18.2). The contract administrator must specify in writing exactly which work may remain, and an appropriate deduction is made from the contract sum (cl 4.14.3). It is advisable that the value of the deduction is agreed before the work is accepted (see *Mul* v *Hutton Construction Limited*). Sometimes when defective work is retained, a variation is needed to other work, in order to accommodate the change. If such a consequential variation becomes necessary (again following consultation with the contractor) no addition is made to the contract sum and no extension of time or direct loss and/or expense is given in respect of this (cl 3.18.3). Where the work is part of the contractor's designed portion, the contract administrator has no authority to allow or require it to remain. If the employer prefers that the work is allowed to stay (perhaps seeking to avoid delays to the programme), this would have to be agreed with the contractor, which would presumably seek to limit its liability for any defects in the non-compliant work.

Mul v *Hutton Construction Limited* [2014] EWHC 1797 (TCC)

This case concerned what constitutes an 'appropriate deduction' when an employer decided to accept non-conforming work. The project concerned an extension and refurbishment work to a country house using the JCT IC05 form. A practical completion certificate was issued with a long list of defects attached, and during the rectification period the employer decided to have this work corrected by other contractors. The employer then started court proceedings against the contractor, to claim back the costs of this work.

A key issue was the interpretation of clause 2.30, which provides that the contract administrator can instruct the contractor not to rectify defects and 'If he does so otherwise instruct, an appropriate deduction shall be made from the Contract Sum in respect of the defects, shrinkages or other faults not made good'. In this case the contractor argued that an 'appropriate deduction' was limited to the relevant amount in the contract rates or priced schedule of works. The court disagreed. It decided that 'appropriate deduction' under clause 2.30 meant 'a deduction which is reasonable in all the circumstances', and could be calculated by any of the following: the contract rates or priced schedule of works; the cost to the contractor of remedying the defect (including the sums to be paid to third party sub-contractors engaged by the contractor); the reasonable cost to the employer of engaging another contractor to remedy the defect; or the particular factual circumstances and/or expert evidence relating to each defect and/or the proposed remedial works.

However, the judge also pointed out that the employer will still have to satisfy the usual principles that apply to a claim for damages, which include showing that it mitigated its loss. If the employer unreasonably refused to let the contractor rectify defects, then it is likely to find its damages limited to what it would have cost the contractor to put them right.

5.53   If work has been shown to be defective, and further similar non-compliance is suspected, then further tests may have to be ordered (cl 3.18.4). The costs of these, including any direct loss and/or expense, are borne by the contractor. If the work is shown to have been in accordance with the contract, the contractor may be entitled to an extension of time (cl 2.29.2.2) and any direct loss and/or expense (cl 4.22.2.2). Clause 3.18.4 refers to a Code of Practice, which is included in the form in Schedule 4. Its purpose is to 'assist in the fair and reasonable operation' of the provisions regarding further testing. It sets out criteria that the contract administrator should consider when deciding whether to instruct further testing, including, for example, the potential consequences of the non-compliance and the standard of supervision of the work by the contractor.

Non-compliance with clause 2.1

5.54   Clause 2.1 requires the contractor to carry out the work 'in a proper and workmanlike manner' and in accordance with the construction phase plan. Clause 3.19 states that in the event of any failure to comply in this respect, the contract administrator may issue instructions requiring compliance, and these will not result in any addition to the contract sum, nor will they entitle the contractor to any extension of time or direct loss and/or expense. The clause empowers the contract administrator to intervene in the contractor's working methods if necessary.

Sub-contracted work

5.55   SBC16 provides for three methods of sub-contracting work: to sub-contractors selected by the contractor, to sub-contractors 'listed' in the contract documents by the employer and to 'named specialists', who may be identified in the contract documents or through an instruction relating to a provisional sum. All of these methods allow for some control over which firms the contractor uses.

Domestic sub-contractors

5.56 Under clause 3.7 the contractor may only sub-contract work, including the design of the contractor's designed portion, with the written consent of the contract administrator. Failure to obtain this would be a default, providing grounds for termination under clause 8.4.1.4. However, as with other approvals, the contract administrator's permission cannot be unreasonably withheld (cl 1.11.1). It is suggested that permission is required for each instance of sub-letting, rather than agreeing to sub-letting in principle. Clause 3.9 states that 'Where considered appropriate, the Contractor shall engage the sub-contractor using the relevant version of the JCT Standard Building Sub-Contract'. While this is not an absolute requirement, it is suggested that the contractor must be able to justify any departure from this policy. JCT Ltd publishes a suite of sub-contracts developed for use with SBC16, including one for use where the sub-contracted work relates to the contractor's designed portion. Whatever form of domestic sub-contract is used, clause 3.9 requires that it must include certain conditions, including that:

- the sub-contract is terminated immediately upon termination of the main contract (cl 3.9.1);

- unfixed materials and goods placed on the site by the sub-contractor shall not be removed without written consent by the contractor (cl 3.9.2.1);

- it shall be accepted that materials or goods included in an interim certificate that have been paid for by the employer become the property of the employer (cl 3.9.2.1.1);

- it shall be accepted that any materials or goods paid for by the main contractor prior to being included in a certificate become the property of the main contractor (cl 3.9.2.1.2);

- the sub-contractor shall provide access for the contract administrator to workshops, etc. (cl 3.9.2.2);

- each party will comply with its obligations under the CDM Regulations (cl 3.9.2.3);

- the sub-contractor has a right to interest on late payments by the contractor at the same rate as that due on main contract payments (cl 3.9.2.4);

- the sub-contractor will grant third party rights and/or enter into warranties as required under the main contract (cl 3.9.2.5);

- the sub-contractor shall provide information and grant licenses reasonably necessary for the contractor to fulfil its obligations under clauses 2.40 and 3.23 and/or as required under the BIM protocol (cl 3.9.3).

5.57 Clause 3.9.2.1 is to protect the position of the employer and the provisions regarding unfixed goods and materials are of particular importance in this respect. If a main contractor should sub-contract on other terms, and this results in losses to the employer, then the contractor may be liable as this would be a breach of contract.

5.58 Once materials have been built in, under common law they would normally become the property of the owner of the land, irrespective of whether or not they have been paid for by the contractor. This would be the case even if there were a retention of title clause in the contract with the sub-contractor or supplier. A retention of title clause is one which stipulates that the goods sold do not become the property of the purchaser until they have been paid for, even if they are in the possession of the purchaser.

5.59   The employer could be at risk, however, where materials have not yet been built in, even where the materials have been certified and paid for. The contractor might not actually own the materials paid for because of a retention of title clause in the sale of materials contract. Under the Sale of Goods Acts 1979, sections 16–19, property in goods normally passes when the purchaser has possession of them, but a retention of title clause will be effective between a supplier and a contractor even where the contractor has been paid for the goods, provided they have not yet been built in. It should be noted, however, that the employer may have some protection through the operation of section 25 of the Act, which in some circumstances allows the employer to treat the contractor as having authority to transfer the title in the goods, even though this may not in fact be the case (*Archivent* v *Strathclyde Regional Council*).

> *Archivent Sales & Developments Ltd* v *Strathclyde Regional Council* (1984) 27 BLR 98 (Court of Session, Outer House)
>
> Archivent agreed to sell a number of ventilators to a contractor which was building a primary school for Strathclyde Regional Council. The contract of sale included the term 'Until payment of the price in full is received by the company, the property in the goods supplied by the company shall not pass to the customer'. The ventilators were delivered and included in a certificate issued under the main contract (JCT63), which was paid. The contractor went into receivership before paying Archivent, who claimed against the Council for the return of the ventilators or a sum representing their value. The Council claimed that section 25(1) of the Sale of Goods Act 1979 operated to give it unimpeachable title. The judge found for the Council. Even though the clause in the sub-contract successfully retained the title for the sub-contractor, the employer was entitled to the benefit of section 25(1) of the Sale of Goods Act 1979. The contractor was in possession of the ventilators and had ostensible authority to pass the title on to the employer, who had purchased them in good faith.

5.60   Another risk relating to rightful ownership is where the contractor fails to pay a domestic sub-contractor who has purchased materials and the sub-contractor claims ownership of the unfixed materials. Here, the risk may be higher, as a work and materials contract is not governed by the Sale of Goods Act 1979. Therefore, there can be no assumption that property would pass on possession.

5.61   SBC16 attempts to deal with the issues surrounding ownership in several ways. First, unfixed materials and goods which have been delivered to the site and intended for the works may not be removed without the written consent of the contract administrator (cl 2.24 and 3.9.2.1). Removal would be a breach of contract, therefore the employer could claim from the contractor for any losses suffered through unauthorised removal. This would apply even though the materials or goods may not yet have been included in any certificate. Second, unfixed materials and goods either on or off site which have been included in an interim payment are to become the property of the employer (cl 2.24 and 2.25), and the contractor is thereby prevented from disputing ownership.

5.62   Clauses 2.24 and 2.25 of the main contract, however, are only binding between the parties, and do not place obligations on any sub-contractor. The risk facing the employer is that if the contractor becomes insolvent, a sub-contractor or supplier may still have a rightful claim to ownership of the unfixed goods, even though they have been paid for by the employer (see *Dawber Williamson Roofing* v *Humberside County Council*). The main contract therefore requires that all sub-contracts include similar clauses to 2.24 and 2.25 regarding non-removal from site and ownership passing upon payment (cl 3.9.2.1). Sub-contracts must also include a clause stating that once materials and goods have been certified and paid for under the main contract they become the property of the employer and that the sub-contractor 'shall not deny' this. This would operate even where the main contractor has become insolvent. Even this provision might not protect the employer in some circumstances because if the sub-contractor does not have 'good title' it cannot pass it on. Thus, for example, it might not prevent a sub-subcontractor claiming ownership.

Dawber Williamson Roofing Ltd v *Humberside County Council* (1979) 14 BLR 70

The plaintiff entered into a sub-contract with Taylor and Coulbeck Ltd (T&C) to supply and fix roofing slates. The main contractor's contract with the defendant was on JCT63. By clause 1 of their sub-contract (which was on DOM/1), the plaintiff was deemed to have notice of all the provisions of the main contract, but it contained no other provisions as to when property was to pass. The plaintiff delivered 16 tons of roofing slates to the site, which were included in an interim certificate, which was paid by the defendant. T&C then went into liquidation without paying the sub-contractor, who brought a claim for the amount or, alternatively, the return of the slates. The judge allowed the claim, holding that clause 14 of JCT63 could only transfer property where the main contractor had a good title. (The difference between this and the Archivent case cited above is that in this case the sub-contract was a contract for work and materials, to which the Sale of Goods Act 1979 did not apply.) Provisions within clause 3.9.2 of SBC16 now deal with the problem illustrated by this case.

Listed sub-contractors

5.63   Under clause 3.8 the contractor's choice of a sub-contractor to carry out certain work, measured or otherwise described, can be restricted to any one of three or more persons named in the contract bills, or in a list annexed to the bills. The contractor must select one of those listed and is responsible for the performance of such sub-contractors to the same degree as it would be for any sub-contractor it had selected itself. All sub-contracts should contain the provisions described above and the contractor may be required to obtain warranties from the listed sub-contractor (see paragraph 2.59).

5.64   Names may be added to the list by either the employer or the contractor, with the consent of the other (cl 3.8.1). If less than three of those listed are able or willing to carry out the work then the employer (or the contract administrator on its behalf) and the contractor may add names to bring the total up to no fewer than three (cl 3.8.1, see Figure 5.2). If it is not possible to maintain a list of three, then the work may be carried out by the contractor, or sub-let to a domestic sub-contractor under clause 3.7 (cl 3.8.2). The contractor has the right of reasonable objection to any new addition to the list, and given that it is taking responsibility for the sub-contractor's performance, it is likely that any real concern about the competence of the firm would be a good reason for withholding its consent. Where there are difficulties in engaging any of those preferred by the employer, the employer may be considered unreasonable if it objects to any alternative suggestions put forward by the contractor.

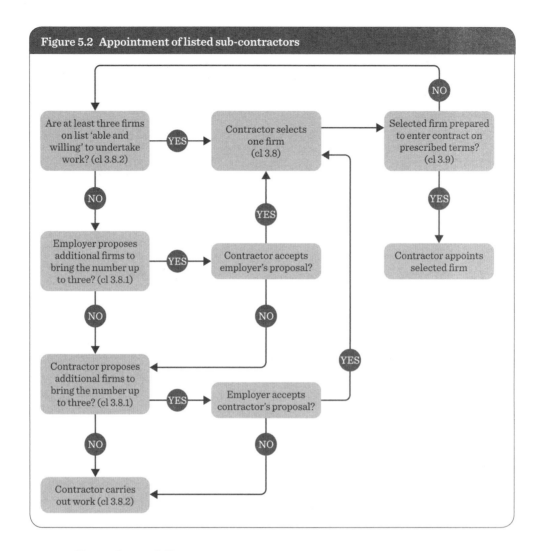

Figure 5.2 Appointment of listed sub-contractors

Named specialists

5.65   The provisions for 'Named Specialists' are set out in Supplemental Provision 9 of Schedule 8. They give the employer the right to require the contractor to use a particular firm to undertake part of the works. CDP work is specifically excluded from this provision (Schedule 8:9.1) and, unlike the more detailed named sub-contractor provisions in the Intermediate Building Contract with Contractor's Design (ICD16), there is no reference to named specialists undertaking a design role in either the contract or in the *JCT Standard Building Contract Guide* (SBC/G). If a design role is intended, parties would need to consider many matters, including whether or not the contractor is to be responsible for that design, the level of liability and whether professional indemnity insurance and collateral warranties are required; all of these would require carefully drafted amendments to the form.

5.66   Despite this limitation, the provision may be useful in situations where the employer wishes to involve particular firms in certain aspects of construction. The employer may, for

Table 5.6 Risk distribution

Risk	Extension of time	Loss and/or expense
Named specialist's progress causes delay	No	No
Instructions relating to provisional sums for post-named specialist work under Schedule 8:9.1.2	Yes (cl 2.29.2.1/3.16)	Yes (cl 4.22.2.1/3.1)
Delay in issuing or failure to issue instructions dealing with naming	Yes (cl 2.29.7)	Yes (cl 2.29)
Instructions required the contractor to undertake a named specialist's work itself under Schedule 8:9.3.3	No	No
Instructions under Schedule 8:9.4.1 where the contractor notifies reasonable objection to a named specialist who was not named in the contract documents	Yes (cl 2.29.2.3)	Yes (cl 4.22.2.4)
Instructions under Schedule 8:9.4.2 where the contractor notifies a reasonable objection to a replacement for a named specialist	Yes (cl 2.29.2.3)	Yes (cl 4.22.2.4)
Named specialist becomes insolvent	Yes (cl 2.29.14)	See below
Instructions issued under Schedule 8:9.6 after a named specialist has become insolvent	Yes (implied)	Yes (cl 4.22.2.5)

example, have used the firm on other projects and found it to be particularly reliable, and/ or its workmanship to be exceptional. The contractor remains entirely responsible for the quality of the firm's work. There are, however, some risks to the employer; these would arise if there were difficulties in entering into a contract with the named specialist, or the firm became insolvent. (The resulting risk distribution is summarised in Table 5.6.)

5.67 If the employer wishes the contractor to engage a named specialist, this can be done in two ways. The first is to name the firm and identify the work in the contract documents (the work is termed 'Pre-Named Specialist Work'); the second is to do this under an instruction relating to a provisional sum (the work is termed 'Post-Named Specialist Work). In the latter case, the contractor may make a reasonable objection to the firm.

5.68 The contractor must enter into an agreement with the named specialist 'as soon as reasonably practicable' (Schedule 8:9.2). The contractor should engage the specialist on terms that comply with clause 3.9, i.e. using the appropriate JCT sub-contract or incorporating the provisions set out in that clause (Schedule 8:9.2; see paragraph 5.56). Other than this, no particular terms are required for the sub-contract, unless the employer has stipulated special conditions in the tender documents or in the instruction.

5.69 If the contractor is unable to enter into a sub-contract it must immediately inform the contract administrator of the reasons that have prevented this from happening (Schedule 8:9.3, see Figure 5.3). Provided the contractor's objections are reasonable, the contract administrator must, within seven days, by an instruction either remove the reason, or select another named specialist to carry out the work, or direct the contractor to undertake the work itself (or have it carried out by a sub-contractor selected by the contractor and approved by the employer), or omit the named specialist work as a variation (Schedule 8:9.3.4). If the named specialist's contract is terminated (see paragraph 9.34), the contract administrator has the same options regarding issuing instructions to remedy the situation.

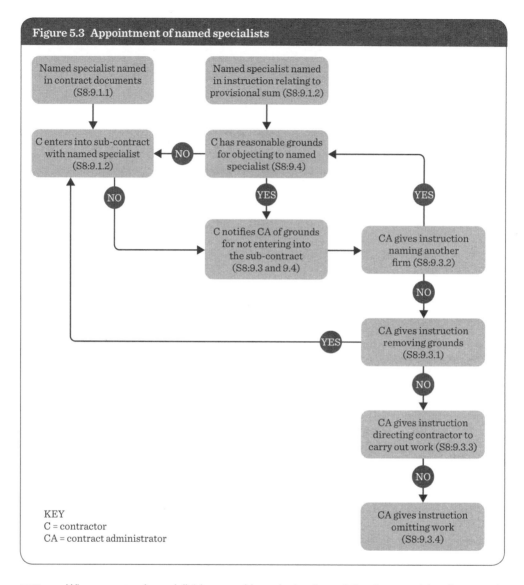

Figure 5.3 Appointment of named specialists

Named specialist named in contract documents (S8:9.1.1)

Named specialist named in instruction relating to provisional sum (S8:9.1.2)

C enters into sub-contract with named specialist (S8:9.1.2)

C has reasonable grounds for objecting to named specialist (S8:9.4) — **NO** →

YES ↓ C notifies CA of grounds for not entering into the sub-contract (S8:9.3 and 9.4)

NO ↑

CA gives instruction naming another firm (S8:9.3.2)

YES

NO

CA gives instruction removing grounds (S8:9.3.1) — **YES**

NO

CA gives instruction directing contractor to carry out work (S8:9.3.3)

NO

CA gives instruction omitting work (S8:9.3.4)

KEY
C = contractor
CA = contract administrator

5.70   Where a named specialist is named in an instruction relating to a provisional sum, or in an instruction to remove difficulties as described above (but was not named in the contract), then this may be grounds for claiming an extension of time or loss and/or expense (cl 2.29.2.3 and 4.22.2.4). If a named specialist becomes insolvent, then this may also provide grounds for such claims (cl 2.29.14 and 4.22.2.5).

Work not forming part of the contract/persons engaged by the employer

5.71   Under clause 2.7 the employer may engage persons directly to carry out work that does not form part of the contract, while the main contractor is still in possession. If the contract bills have included this requirement, then the contractor must permit the employer to execute such work (cl 2.7.1). Otherwise, the employer can only do this with the contractor's permission (cl 2.7.2). The permission may not be unreasonably delayed or withheld.

5.72   It should be remembered that the indemnities given by the contractor under clauses 6.1 and 6.2 (and therefore the insurances under clause 6.4) do not extend to persons employed under clause 2.7. The employer should also be made aware that any serious disruption caused to the contractor's working could lead to an extension of time (cl 2.29.7), to loss and/or expense (cl 4.22.5), or even to termination (cl 8.9.2.2).

Making good defects

5.73   The contract administrator may instruct the contractor to make good any 'defects, shrinkages or other faults' which appear during the rectification period (cl 2.38.2). The power is limited to those defects that result from the work not having been carried out in accordance with the contract or to the contractor failing to comply with its obligations with respect to the contractor's designed portion. This does not include other defects that may be due, for example, to errors in the design information supplied to the contractor, or to general wear and tear resulting from occupation by the employer. The power is also limited to those defects that appear after practical completion, although it would be sensible to allow the contractor to correct any outstanding defects (*Pearce and High* v *Baxter*). The power would cover defects appearing during the rectification period that have been caused by frost, but only where the damage was due to a default of the contractor before practical completion.

> *Pearce and High* v *John P Baxter and Mrs A S Baxter* [1999] BLR 101 (CA)
>
> The Baxters employed Pearce and High on MW80 to carry out certain works at their home in Farringdon. Following practical completion, the architect issued interim certificate no. 5, which the employer did not pay. The contractor commenced proceedings in Oxford County Court, claiming payment of that certificate and additional sums. The employer, in its defence and counterclaim, relied on various defects in the work that had been carried out. Although the defects liability period had by that time expired, neither the architect nor the employer had notified the contractor of the defects. The Recorder held that clause 2.5 was a condition precedent to the recovery of damages by the employer, and further stated that it was a condition precedent that the building owner has notified the contractor of patent defects within the defects liability period. The employer appealed and the appeal was allowed. Lord Justice Evans stated that there were no clear express provisions within the contract which prevented the employer from bringing a claim for defective work, regardless of whether notification had been given. He went on to state, however, that the contractor would not be liable for the full cost to the employer of remedying the defects, if the contractor had been effectively denied the right to return and remedy the defects itself.

5.74   If an instruction to make good defects is issued, the contractor must comply 'within a reasonable time' (cl 2.38). Unless causing unacceptable problems, defects that appear during the rectification period are normally left until after the expiration of the period. The power to instruct ends at the time the schedule of defects is issued or 14 days after the end of the rectification period.

5.75   The contract administrator has a duty under clause 2.38.1 to issue a schedule of defects not later than 14 days after the end of the rectification period, the only point where the contract requires the contract administrator to issue such a schedule. The schedule is issued in the form of an instruction, and the contractor is required to rectify the defects within a reasonable time. If the contract administrator, with the agreement of the employer, decides to accept the defective work then this should be clearly shown in the instruction, and under clause 2.38 an appropriate deduction is made from the contract sum (cl 4.14.3). Care should be taken to establish the full extent of the problem before such a course of action is taken, as it is unlikely that the employer would thereafter be able to claim for consequential problems or further remedial work. In particular, if the work was stated to be 'to approval' in the contract documents, this will have implications for the final certificate (see paragraphs 7.48–7.51) which should be resolved before the work is accepted.

5.76   Once satisfied that all the defects have been made good, the contract administrator must issue a certificate to that effect (a 'Certificate of Making Good') (cl 2.39). The certificate is one of the preconditions to the issue of the final certificate. The contract does not state what should happen in respect of defects which appear after the issue of the certificate but before the issue of the final certificate. It is, however, clear from clause 2.38 that the contract administrator no longer has the power to instruct that these are made good. It is suggested that in such circumstances there would be two possible courses of action. The first would be to make an agreement with the contractor to rectify the defects before the final certificate is issued. If the contractor refuses to do this, an amount could be deducted from the contract sum to cover the cost of making good the work, but this would involve some risk to the employer. The second and less risky course would be to have the defective work rectified by another contractor, and to deduct the amount paid from the contract sum. This would involve a delay to the issue of the final certificate and would probably be disputed by the contractor.

6 Sums properly due

6.1 SBC16 With Quantities and Without Quantities are both lump sum contracts. If the 'with quantities' version is used, the work will be set out in bills of quantities prepared using the 'Measurement Rules' – the RICS *New Rules of Measurement 2: Detailed Measurement for Building Works* (NRM2), unless otherwise stated in the bills. In the 'without quantities' version there will be no bills, but the contractor may be required to price a specification or schedule of works (pricing option A) or may supply a contract sum analysis or schedule of rates (pricing option B). In both cases, the contract sum will be the tender figure accepted or agreed following negotiation and is entered in Article 2. However, this is rarely the amount actually paid. The wording of the contract recognises this by the qualifying reference 'or such other sum as becomes payable' (Article 2).

6.2 The contract sum may contain provisional sums or approximate quantities to cover the cost of work that cannot be accurately described or measured until work is under way. Most contracts will require some variations to the works. There is also the possibility of claims from the contractor for loss and/or expense arising from intervening events that could not be foreseen at the time of tendering. While it is possible in theory to make a contract 'fixed price', most will allow for fluctuations to some degree. Fees or charges in respect of statutory matters which are not allowed for in the contract bills will require an adjustment to the sum. Under the supplemental provisions, the contractor is encouraged to propose cost-saving and value-improvement measures, which may also result in a change (Schedule 8, Supplemental Provision 3). In addition, if an amount is agreed following an acceleration quotation (Schedule 2), then this is to be added to the contract sum.

6.3 There will therefore almost inevitably be adjustments to the contract sum; SBC16 clause 4.2, however, makes it clear that the only alterations that may be made are those provided for in the terms. Items to be included in adjustments are set out in clause 4.3, and the ascertained amounts will be added or deducted as appropriate at the next interim certificate (cl 4.4).

6.4 Arithmetical errors by the contractor in pricing are not allowed as a cause for adjustment. Errors in the preparation of the contract bills, on the other hand, must be corrected and will then be treated as if they are a variation (cl 2.14.1 and 2.14.3). Any divergence between the contract drawings and other documents which necessitates an instruction by the contract administrator may also result in a variation (cl 2.15).

6.5 SBC16 With Approximate Quantities is a remeasurement contract. Approximate quantities only are given for all of the work, and the contractor submits a fully priced copy of the bills of approximate quantities at tender stage, which forms the basis of the contract. No 'Contract Sum' is entered in the articles. All the work is remeasured prior to certification and the contractor is paid for the actual quantities of work carried out. The final amount payable in accordance with the conditions is termed the 'Ascertained Final Sum' (Article 2).

An approximate quantity

6.6   Under SBC16 With Quantities, where work can be described in accordance with the NRM2, but where the quantity involved is uncertain, an 'Approximate Quantity' can be included in the bills. The contract administrator is not required to issue any further instruction for the contractor to carry out this work. After it has been carried out, the work is valued using the rate or price given for the approximate quantity (cl 5.6.1.4).

6.7   Difficulties can arise if the approximate quantity is not a reasonably accurate forecast of the quantity of work required. In such a situation, the valuation must include a fair allowance for the difference in quantity over and above the rates or prices tendered by the contractor (cl 5.6.1.5). The contractor may also claim that the inaccuracy is a 'Relevant Event' under clause 2.29.5 and a 'Relevant Matter' for direct loss or expense under clause 4.22.4. Similar valuation rules are set out in the With Approximate Quantities version of SBC16.

Provisional sums

6.8   If insufficient information can be provided at the time of tender to allow an item to be described and measured in accordance with the NRM2, then a provisional sum may be inserted in the bills to cover the item (NRM2 Rule 2.9.1). The contract administrator must issue instructions regarding all work covered by provisional sums in the contract bills and the contractor can take no action with regard to this work until it has received an instruction (cl 3.16). Under NRM2, provisional sums are either for defined work or for undefined work. Provisional sums may also be included in the specification/work schedules in the Without Quantities version. These are dealt with in the same way as those for undefined work, described below. In all cases, the work covered by a provisional sum is valued by the quantity surveyor.

Defined work

6.9   The information required to place provisional work in the defined category is listed in NRM2 Rule 2.9.1.2. The tenderer must be aware of the nature and construction of the work, how and where the work fits into the building, the scope and extent of the work and any specific limitations on method, sequence or timing. In other words, the description must be sufficiently detailed for the contractor to make proper allowance for the effect of the work when pricing the relevant preliminaries, and to allow for the work in the programme.

6.10   If the information provided is not as detailed as the rule requires, or if it is erroneous, then a corrective instruction is required from the contract administrator (cl 2.14.1). This will be treated as a variation (cl 2.14.3) and could give rise to a notice of delay (cl 2.29.1) and an application for reimbursement of direct loss and/or expense (cl 4.22.1) from the contractor. The corrective action cannot simply be to change from the defined category to the undefined category by substituting a new provisional sum.

Undefined work

6.11   A provisional sum for undefined work will be applicable where it is not possible to supply the amount of information needed to comply with NRM2 Rule 2.9.1.2. The contractor will not have been able to make proper allowance for the work in its programming and planning or the pricing of preliminaries. A provisional sum in this case should be sufficient not only to cover the net cost, but also to take into account the fact that there might be additions to preliminaries, attendance and perhaps loss and/or expense, etc. There is also a risk that the contractor might give notice of delay arising from the contract administrator's instruction. Unlike an instruction for the expenditure of a provisional sum for defined work, that for a provisional sum for undefined work could be a relevant event (cl 2.29.2.1) and a relevant matter for which a loss and/or expense application can be made (cl 4.22.2.1).

6.12   Provisional sums may be included for items that are not specifically work (e.g. testing, site boards, site facilities). The heading in the bills of quantities will simply be 'include the Provisional Sum of _____ for ____', or some other appropriate wording. For work to be carried out by statutory authorities it is suggested that the description of the work be followed by a similar heading.

Valuation of variations

6.13   There are three mechanisms by which a variation can be valued under the provisions of the contract (see Figure 6.1). Clause 5.2 requires that all variations and all instructions relating to the expenditure of provisional sums are, unless agreed between the parties, valued by the quantity surveyor using the 'Valuation Rules' in accordance with the provisions of clauses 5.6 to 5.10 (see paragraphs 6.19–6.26). The second mechanism is through the submission of a 'Variation Quotation' (cl 5.3 and Schedule 2). In both cases, the valuation should include a sum in respect of any additional design work required by the instruction. Under Supplemental Provisions 3 and 4, the contractor is encouraged to propose cost savings, value improvements and environmental performance improvement measures. The value of any resultant variation will be a matter for agreement between the parties.

Schedule 2 variation quotations

6.14   If the employer or the contract administrator wishes to ascertain the contractor's price for a variation, then the instruction requiring the variation should request that a quotation is submitted in accordance with Schedule 2 (cl 5.3.1). The contractor has seven days to object to the application of this procedure. If the contractor objects, the instruction is not carried out unless the contract administrator instructs that it should be, in which case it is valued 'by a Valuation', i.e. by the quantity surveyor (cl 5.3.2). The instruction should include sufficient detail to enable the contractor to provide the information required. Any addendum to the contract bills issued for the purposes of obtaining a Schedule 2 quotation should be prepared according to the measurement rules unless otherwise stated (cl 2.13.1). Any errors in the addendum are to be treated in the same way as errors in the contract bills (cl 2.14.1).

6.15   If no objection is raised, the contractor must submit the Schedule 2 quotation within 21 days of receipt of the instruction or any additional information requested under Schedule 2:1.1. The quotation should identify the direct cost of complying with the instruction, the period required for extension to the contract period and the sum acceptable in lieu of direct loss and/or expense. The quotation should make reference where relevant to rates and prices in the contract bills (Schedule 2:2.1).

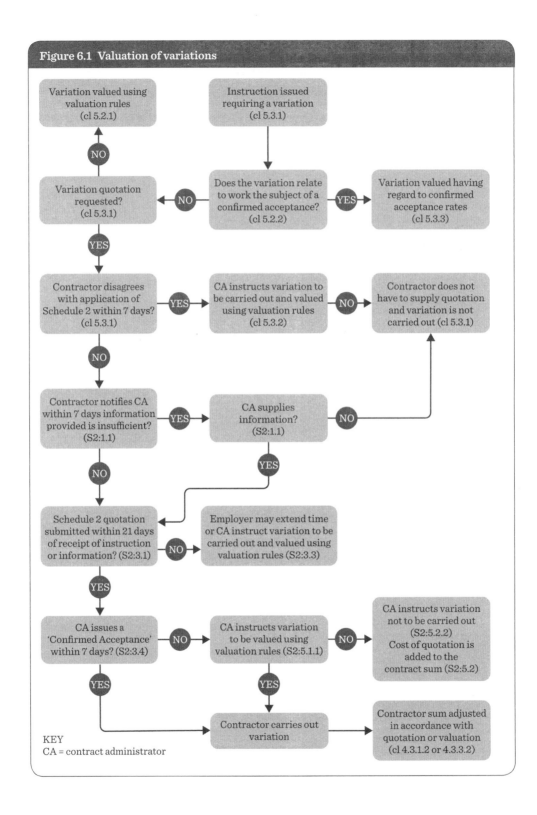

Figure 6.1 Valuation of variations

6.16   If accepted, the quotation takes the place of valuation by the quantity surveyor. This method brings certainty of outcome for the parties, as both are bound by what is agreed with respect to the value of the work, the extension of time and the direct loss and/or expense (for fluctuations, see paragraph 6.45). The certainty, however, is likely to be secured only at a price, particularly where the variation does not relate to work for which there are rates and prices in the bills. If the quotation is rejected, the work can still be instructed, but it is then subject to valuation by the quantity surveyor (Schedule 2:5.1). The contractor is paid a fair and reasonable amount for the cost of preparing the Schedule 2 quotation (Schedule 2:5.2).

6.17   If the contract administrator subsequently issues a variation to work for which a Schedule 2 quotation has been given and accepted, then this variation is valued on a fair and reasonable basis by the quantity surveyor, 'having regard to' the contents of the original Schedule 2 quotation (cl 5.3.3). The clause states that the valuation rules are only applicable 'to the extent that they are consistent with those requirements'. In summary, the parties will be bound by the terms agreed in the original Schedule 2 quotation with respect to both the original instruction and to any future related variations.

6.18   It should be noted that the Schedule 2 provisions do not appear to apply to instructions regarding the expenditure of provisional sums, as clause 5.3.1 refers only to 'a Variation'. However, there would be nothing to prevent the contract administrator or employer requesting a quotation prior to issuing such an instruction.

Valuation by the quantity surveyor

6.19   If no Schedule 2 quotation is sought, or if the quotation is rejected, and the valuation is not otherwise agreed between the employer and contractor, then the valuation of the variation must be made by the quantity surveyor (termed a 'Valuation') according to the rules set out in clause 5.6.1 (cl 5.2.1). If the variation involves an omission then the value of the work shown in the contract bills is deducted from the contract sum (cl 5.6.2).

6.20   Clause 5.6.1 (With Quantities version) includes for:

*   work of similar character undertaken under similar conditions and where the quantity does not change significantly (cl 5.6.1.1);

*   work of similar character but not undertaken under similar conditions and/or where the quantity changes significantly (cl 5.6.1.2);

*   work not of similar character (cl 5.6.1.3).

6.21   In the first two cases, the rates and prices in the bills of quantities are to be used in assessing the value of the variation; it should also be noted that the work is not necessarily identical, and that the contractual rates must be used even where those figures contain errors (*Henry Boot Construction Ltd* v *Alstom Combined Cycles*). In the third case, the work should be valued at 'fair rates and prices'. Dissimilar conditions might include, for example, that the instructed work is carried out in winter, whereas under the bills it had been assumed it would be carried out in summer. Such an assumption, however, would have to be clear from an objective analysis of the contract documents (*Wates Construction* v *Bredero Fleet*).

Henry Boot Construction Ltd v Alstom Combined Cycles [2000] BLR 247

By a contract formed in 1994, Alstom Combined Cycles employed Henry Boot to carry out civil engineering works in connection with a combined cycle gas turbine power station for PowerGen plc at Connah's Quay in Clwyd. During post-tender negotiations, a price of £258,850 was agreed for temporary sheet piling to trench excavations. Disputes arose regarding the valuation of this work, and these disputes were initially taken to arbitration. The arbitrator found that the agreed figure contained errors that effectively benefited Boot. Boot argued that, nevertheless, the figure should be used to value the work under clause 52(1). The arbitrator decided that 52(1) (a) and (b) were inapplicable, and that 52(2) should be applied to achieve a fair valuation. Boot appealed to the Technology and Construction Court, and Judge Humphrey Lloyd decided that the mistake made no difference; the agreed rate should be used even if the results were unreasonable. Clause 52(2) created only a limited exception where the scale or nature of the variation itself made it unreasonable to use the contract rates.

Wates Construction (South) Ltd v Bredero Fleet Ltd (1993) 63 BLR 128

Wates Construction entered into a contract on JCT80 to build a shopping centre for Bredero. Some sub-structural work differed from that shown on the drawings and disputes arose regarding the valuation of the works, which were taken to arbitration. In establishing the conditions under which, according to the contract, it had been assumed that the work would be carried out, the arbitrator took into account pre-tender negotiations and the actual knowledge that Wates gained as a result of the negotiations, including proposals that had been put forward at that time. Wates appealed and the court found that the arbitrator had erred by taking this extrinsic information into consideration. The conditions under which the works had to be executed had to be derived from the express provisions of the bills, drawings and other contract documents.

6.22   Clause 5.7 (daywork) provides for work which cannot properly be valued by measurement. In such cases the valuation is based on the prime cost of the work, calculated in accordance with the definitions of prime cost referred to in clauses 5.7.1 and 5.7.2. The contractor must provide vouchers for verification by the contract administrator, showing specified details of the daywork no later than seven business days after the work is carried out.

6.23   Where, as a result of a variation, other contract work has to be carried out under different conditions, then this must be treated as if it were a variation and valued accordingly, even though the consequences were not themselves identified in the original instruction (cl 5.9).

6.24   In all cases, any measurement should be made according to the principles governing the preparation of the contract bills (i.e. NRM2 unless otherwise stated therein, see clauses 2.13.1 and 5.6.3.1). Clause 5.6.3.3 authorises appropriate allowance to be made for an addition to or reduction of preliminaries, except for instructions regarding the expenditure of provisional sums for defined work.

6.25   In the Without Quantities version, the clause 5.6.1 valuation rules are slightly simpler. Work of a similar character is to be valued according to rates and prices in the priced document, with a fair allowance being made if there is any change in conditions under which the work is carried out, or any significant change in quantity. Where the work is not of similar character it should be valued at fair rates and prices.

6.26 The value of variations to the contractor's designed portion 'shall be consistent with the values of work of a similar character set out in the CDP Analysis, making due allowance for any change in the conditions under which work is carried out … Where there is no work of a similar character … a fair valuation shall be made' (cl 5.8.2). The rules under clauses 5.6.3, 5.7 and 5.9 apply as relevant.

Reimbursement of direct loss and/or expense

6.27 The objective of the 'Loss and Expense' provisions is to enable the contractor to be reimbursed for direct loss and/or direct expense suffered as a result of delay or disruption and for which the contractor is not reimbursed under any other provision in the contract. The contractor is entitled to be reimbursed for loss and/or expense incurred as a result of deferment of possession or any occurrence of a 'Relevant Matter' set out in clause 4.22 (cl 4.20). The amount to be paid is determined under the procedure in clause 4.21 or through acceptance of a variation quotation or an acceleration quotation (Schedule 2).

6.28 Under clause 4.20, the entitlement to direct loss and/or expense is subject to the contractor having complied with the procedure set out in clause 4.21 (cl 4.20.1). This requires the contractor to notify the contract administrator promptly, i.e. 'as soon as the likely effect of a Relevant Matter on regular progress or the likely nature and extent of any loss and/or expense arising from a deferment of possession becomes (or should have become) reasonably apparent to him' (cl 4.21.1). The notice is to be accompanied by, or followed by, an assessment of the losses already incurred and those likely to be incurred (cl 4.21.2). The contractor must keep the contract administrator updated at monthly intervals until all information reasonably required and necessary for ascertaining the amount due has been supplied to the contract administrator (cl 4.21.3)

6.29 The contract administrator or quantity surveyor is required to notify the contractor of the ascertained amount of loss and/or expense within 28 days of receipt of the initial assessment and information, and subsequently within 14 days of receipt of each monthly update of the assessment and information (cl 4.21.4). Each ascertainment must be made by reference to the information supplied by the contractor and be in sufficient detail to allow the contractor to identify differences between its own assessment and the contract administrator's ascertainment.

6.30 The procedure is more detailed and contains stricter time limits than that in SBC11. It ensures that the contract administrator is kept fully up to date with the effect and likely costs associated with any relevant event. As well as allowing the employer to budget for the additional costs, there may be steps that can be taken at an early stage to minimise the potential increase. It also ensures that the contractor is informed at an early date of any disagreement by the contract administrator with the contractor's assessment, and is updated on a regular basis as to any changes in that position.

6.31 Importantly, as noted above, the contractor's right to loss and expense is 'subject to … compliance with the provisions of clause 4.21'. The courts held even on earlier, less clear versions of this clause that the right to loss and/or expense could be lost if the contractor did not act promptly (see *London Borough of Merton* v *Leach*). Given the new wording, there is no doubt that the employer could refuse to consider late applications. However, it is still arguable that for matters that would constitute a breach of contract on behalf of the employer, the contractor might retain the right to claim damages under common law (a right confirmed by clause 4.24). Therefore it may be sensible to agree that such claims

London Borough of Merton v *Stanley Hugh Leach Ltd* (1985) 32 BLR 51 (ChD)

Stanley Hugh Leach entered into a contract on JCT63 with the London Borough of Merton to construct 287 dwellings. The contract was substantially delayed and a dispute arose regarding this delay and related claims for loss and expense. The dispute went to arbitration and the arbitrator made an interim award on a list of matters. Merton appealed and the court considered 15 questions framed as preliminary issues. Among other things, the court stated that applications for direct loss and/or expense must be made in sufficient detail to enable the architect to form an opinion as to whether there is, in fact, any loss and/or expense to be ascertained. If there is, then it is the responsibility of the architect to obtain enough information to reach a decision. This responsibility could, of course, include requests for information from the contractor. The court also held that the application must be made within a reasonable time and not so late that the architect was no longer able to form an opinion on matters relevant to the application.

should be dealt with under the contract, particularly in cases where the procedural failing on the part of the contractor is minor.

6.32   Not only must the contractor apply promptly, but the applications must be dealt with according to the procedures of the contract, including responding within the tight time limits. Failure to certify an amount properly due will not prevent recovery, and could leave the employer liable in damages for breach of contract (*Croudace* v *London Borough of Lambeth*). Where the contract administrator delegates the duty of ascertaining the direct loss and expense, it appears that it is not obligatory for the contract administrator to accept the quantity surveyor's opinion (*R Burden* v *Swansea Corporation*), although the quantity surveyor's assessment would be strong evidence as to what the correct amount should be. At final account stage there is a requirement to review the awards already made, and to notify the contractor of any further amounts due (cl 4.25.2.1, see paragraph 7.42).

Croudace Ltd v *The London Borough of Lambeth* (1986) 33 BLR 20 (CA)

Croudace entered into an agreement with the London Borough of Lambeth to erect 148 dwelling houses, some shops and a hall. The contract was on JCT63 and the contract administrator was Lambeth's chief architect and the quantity surveyor was its chief quantity surveyor. The contract administrator delegated his duties to a private firm of architects. Croudace alleged that there had been delays and that it had suffered direct loss and/or expense and sent letters detailing the matters to the architects. In reply, the architects told Croudace that they had been instructed by Lambeth that all payments relating to 'loss and expense' had to be approved by the borough. The chief architect then retired and was not immediately replaced. There were considerable delays pending a further appointment and Croudace began legal proceedings. The High Court found that Lambeth was in breach of contract in failing to take the necessary steps to ensure that the claim was dealt with, and was liable to Croudace for this breach. The Court of Appeal upheld this finding.

R Burden Ltd v *Swansea Corporation* [1957] 3 All ER 243 (HL)

Burden entered into a contract with Swansea Corporation to build a school. The contract provided for interim certificates to be issued at intervals by the contract administrator. The contract administrator, who was the Corporation's Borough Architect, acted originally as both contract administrator and surveyor under the contract. Later, after 20 certificates had been issued, the firm of quantity surveyors which had originally prepared the bills was appointed to act as surveyor under the contract in place of the Borough Architect. In the next certificate the surveyor reduced the amount applied for by the contractor by around 75 per cent, and the contract administrator certified the lower figure. The surveyor later discovered that it had made a mistake, but did not inform the contract administrator of the error. The contractor gave notice determining the contract, on the grounds that the employer had interfered with the issue of the certificate. The House of Lords decided that a mistake in a direction as to the amount to be paid did not amount to interference or obstruction. It was suggested that the contract administrator would have been at liberty to have certified a different amount if aware of the error.

6.33   The matters listed in clause 4.22 are concerned with situations where the loss and/or expense is attributable to actions of the contract administrator or the employer, including 'any impediment, prevention or default, whether by act or omission, by the Employer', and excluding the neutral causes which feature in the extension of time provisions of clause 2.29. It should be noted that costs and expenses resulting from the contractor exercising its right to suspend work under clause 4.13 are not dealt with under clause 4.20, but are treated separately (see paragraph 7.22).

6.34   Clause 4.20 refers to regular progress of the works being 'materially affected' by the relevant matter. This could include situations where an overall delay to the programme is experienced (often termed 'prolongation') for which an extension of time may have been awarded. However, it can also include disruption to the planned sequence that does not cause any overall delay, provided it can be shown that losses were suffered as a result. Any disruption claim should be related to the progress necessary to complete the works by the completion date, not necessarily the actual sequences of events on site, and the disruption would have to be significant for it to entitle the contractor to compensation.

6.35   In ascertaining the loss and expense, the contract administrator must determine what has actually been suffered. The sums that can be awarded can include any loss or expense that has arisen directly as the result of the relevant matter. The loss and expense award is in effect an award of damages, and the contract administrator should approach its assessment using the same principles as a court would in awarding damages for breach of contract. In broad terms, the object is to put the contractor back into the position in which it would have been but for the disturbance. The contractor ought to be able to show that it has taken reasonable steps to mitigate its loss, and the losses must have been reasonably foreseeable as likely to result from the 'matter' when the contract was entered into.

6.36   The following are items which could be included:

- increased preliminaries;
- overheads;
- loss of profit;
- uneconomic working;
- increases due to inflation; and
- interest or finance charges.

6.37   The items claimed must not be recoverable by the contractor under any other term of the contract (and, for example, duplication of a claim under clause 5.2 must be avoided).

Prolongation costs, such as on-site overheads, would normally only be claimable for periods following the date for completion. (For head office overheads, etc., see *McAlpine v Property and Land Contractors* below.) Interest may also be recoverable, but only if it can be proved to have been a genuine loss (*FG Minter* v *WHTSO*). As clause 4.20.1 refers to losses which the contractor 'incurs or is likely to incur', the award need not be restricted to losses suffered prior to the time the contractor's application is made, but could include those suffered up to the date of the ascertainment (this would apply particularly to financing charges), and could arguably be extended to losses that could be predicted as likely to occur up to the date the reimbursement is made.

Alfred McAlpine Homes North Ltd v *Property and Land Contractors Ltd* (1995) 76 BLR 59

An appeal arose on a question of law arising out of an arbitrator's award regarding the basis for awarding direct loss and expense with respect to additional overheads and hire of small plant, following an instruction to postpone the works. The judgment contains useful guidance on the basis for awarding direct loss and expense. To 'ascertain' means to 'find out for certain'. It is not necessary to differentiate between 'loss' and 'expense' in a head of claim. Regarding overheads, a contractor would normally be entitled to recover as a 'loss' the shortfall in the contribution that the volume of work had been expected to make to the fixed head office overheads, but which, because of a reduction in volume and revenue caused by the prolongation, was not in fact realised. The fact that 'Emden' and 'Hudson' formulae depend on certain assumptions means that they are frequently inappropriate. The losses on the plant should be the true cost to the contractor, not based on notional or assumed hire charges.

F G Minter Ltd v *Welsh Health Technical Services Organisation* (1980) 13 BLR 1 (CA)

Minter was employed by Welsh Health Technical Services Organisation (WHTSO) under JCT63 to construct the University Hospital of Wales (second phase) Teaching Hospital. During the course of the contract several variations were made and the progress of the works was impeded by the lack of necessary drawings and information. The contractor was paid sums in respect of direct loss and/or expense, but the amounts paid were challenged as insufficient. The amounts had not been certified and paid until long after the losses had been incurred, therefore the figures should have included an allowance in respect of finance charges or interest. Following arbitration, several questions were put to the High Court, including whether Minter was entitled to finance charges in respect of any of the following periods:

(a) between the loss and/or expense being incurred and the making of a written application for the same;
(b) during the ascertainment of the amount; and/or
(c) between the time of such ascertainment and the issue of the certificate including the ascertained amount.

The court answered 'no' to all three questions and Minter appealed. The Court of Appeal ruled that the answer was 'yes' to the first question and 'no' to the others.

6.38　　As noted above, the contractor must provide to the contract administrator all information necessary for ascertaining the amount due (cl 4.21.3), which would normally include full details and particulars of all items concerned with the alleged loss or expense. These should identify which of the losses claimed relate to each of the 'Relevant Matters' that have occurred. This is sometimes compromised by the use of a 'rolled up' or composite claim approach, where it is not really practicable to separate and itemise the effect of a number of causes. This has been accepted by the courts, provided that as much detail as possible has been given, and provided that all disturbance was due to matters under clause 4.22, and not caused by the contractor.

6.39   Formulae such as the 'Hudson' or 'Emden' formulae are sometimes used for estimating head office overheads and profit, which may be difficult to substantiate. Such formulae can be used only where it has been established that there has been a loss of this nature. To do this the contractor must be able to show that, but for the delay, the contractor would have been able to earn the amounts claimed on another contract, for example by producing evidence such as invitations to tender which were declined. Such formulae may be useful where it is difficult to quantify the amount of the alleged loss, provided a check is made that the assumptions on which the formula is based are appropriate.

6.40   Although direct loss and/or expense is a matter of money and not time, which are quite separate issues, there is often a practical correlation in the case of prolongation. Any general implication that there is a link would be incorrect and, in principle, disruption claims and delay to progress are independent. An extension of time, for example, is not a condition precedent to the award of direct loss and/or expense (*H Fairweather & Co.* v *Wandsworth*, see paragraph 4.37).

Fluctuations

6.41   In some projects it may be advantageous to insist on a 'fixed' or 'guaranteed' price, whereby the contractor accepts the risk of all changes in the cost of the works due to statutory revisions and market price fluctuations. However, pricing with this degree of certainty will, in some economic climates, result in higher tender figures, as the contractor will need to allow for possible increases, particularly if the contract period is relatively lengthy. In order to avoid inflated tenders, most contracts allow for some 'fluctuations', whereby the employer accepts some of these risks.

6.42   In SBC16 the default fluctuations provisions are set out in Schedule 7, which allows for contribution, levy and tax fluctuations (Option A). The traditional full fluctuations in labour and materials (Option B) and the use of price adjustment formulae (Option C) are no longer included in the contract, but are available from the JCT website and are referred to in the form. The contract particulars (cl. 4.3 and 4.14) set out the three options, and also allow the parties to state that there will be no fluctuations, or to set out their own provisions; if no selection is made then Option A (i.e. so-called 'fixed price') applies.

6.43   Option A provides for full recovery of all fluctuations in the rates of contributions, levies and taxes in the employment of labour, and in the rates of duties and taxes on the procurement of materials. In short, the only amounts payable are those arising out of an Act of Parliament or delegated legislation. Option B allows, in addition, for fluctuations in the actual market costs of labour and materials. However, contractors have pointed out that many less obvious increases are not included; therefore a 'percentage addition' is made to allow for these. The agreed percentage is entered in the contract particulars. Option C allows for adjustment based on the use of formulae: it does not necessarily take account of the actual costs, but is relatively simple to operate, and is generally considered by contractors to be a fair adjustment.

6.44   Where a contract includes for fluctuations, they will, in the absence of anything to the contrary, be payable for the whole time the contractor is on site, even if it fails to complete within the contract period (*Peak Construction* v *McKinney Foundations*). There is a so-called 'freezing' provision in Schedule 7 of SBC16 (paragraphs A.9, B.10 and C.5), but this depends on the clauses relating to extensions of time being left unamended and all notices of delay being properly dealt with by the contract administrator.

Peak Construction (Liverpool) Ltd v *McKinney Foundations Ltd* (1970) 1 BLR 111 (CA)

Peak Construction was the main contractor on a contract to construct a multi-storey block of flats for Liverpool Corporation. As a result of defective work by nominated sub-contractor, McKinney Foundations, work on the main contract was halted for 58 weeks, and the main contractor brought a claim against the sub-contractor for damages. The Official Referee, at first instance, found that the entire 58 weeks was delay caused by the nominated sub-contractor, and awarded £40,000 of damages, £10,000 of which was for rises in wage rates during the period. McKinney appealed, and the Court of Appeal found that the award of £10,000 could not be upheld as clause 27 of the main contract entitled Peak Construction to claim this from Liverpool Corporation right up until the time when the work was halted.

6.45   The fluctuations provisions also apply in respect of work for which there is a confirmed acceptance of a variation quotation, and to variations to such work, where a base date is set out in the quotation (cl 4.14.1). It would be open to the parties to agree otherwise, provided this is set out in the contract documents. The parties should bear this in mind when dealing with variation quotations, and where a truly fixed figure is desired should make it clear that fluctuations will not apply to the value of the relevant work.

7 Payment

7.1   One of the most important duties of the contract administrator under SBC16, and one to be carried out with care, is the issuing of certificates of payment. It is not unheard of for contractors to become insolvent during the course of a contract, and if the certificates have been overvalued the employer may suffer losses that could have been avoided. On the other hand, the contractor has a right to be paid what the terms of the contract state is due, and the contract administrator must not be influenced by any attempt on the part of the employer to delay certification or withhold amounts properly due.

7.2   Failure to certify correctly could amount to negligence. In the case of *Sutcliffe* v *Thackrah* the House of Lords reversed its own previous judgment, and disposed of the myth that in the role of certifier the contract administrator is operating in a 'quasi judicial' role and is immune from suit. The position is less clear with respect to the contract administrator's duty of care to the contractor. A certifier was found liable to a contractor in the case of *Michael Salliss & Co.* v *Calil and W F Newman & Associates*. Although this appeared to be overtaken in *Pacific Associates Inc.* v *Baxter*, the latter involved non-standard clauses which, if they had not been present, might have resulted in a different outcome.

> *Sutcliffe* v *Thackrah* (1974) 4 BLR 16 (CA)
>
> A contract administrator issued certificates on a contract for the construction of a dwelling house. The contractor's employment was determined for proper reasons, following which the contractor went bankrupt. It then became apparent that much of the work, which had been included in the interim certificates, was defective, and the contract administrator was found negligent. In the House of Lords, when reviewing the role of the contract administrator, Lord Reid stated (at page 21):
>
> > Many matters may arise in the course of the execution of a building contract where a decision has to be made which will affect the amount of money which the contractor gets ... the building owner and the contractor make their contract on the understanding that in all such matters the architect will act in a fair and unbiased manner and it must therefore be implied into the owner's contract with the architect that he shall not only exercise due skill and care but also reach such decisions fairly holding the balance between his client and the contractor.

> *Michael Salliss & Co. Ltd* v *Calil and William F Newman & Associates* (1987) 13 Con LR 69
>
> Calil employed contractor Michael Salliss for some refurbishment works on JCT63. W F Newman acted as contract administrator and quantity surveyor under the contract. The contractor commenced proceedings against the employer and joined the contract administrator as second defendants, claiming that the contract administrator was in breach of its duty to use all professional skill and care in granting only a 12-week extension of time when a 29-week extension was due. There was a sub-trial as to whether the contractor could recover damages

against the contract administrator. His Honour Judge Fox-Andrews held that under a JCT contract the contract administrator owed a duty to the contractor to act fairly between the employer and contractor in matters such as certification and extensions of time. He also noted that (at page 79):

> in many respects a contract administrator in circumstances such as these owes no duty to the contractors. He owes no duty of care to contractors in respect of the preparation of plans and specifications or in deciding matters such as whether or not he should cause a survey to be carried out. He owes no duty of care to a contractor [in deciding] whether or not he should order a variation. Once, however, he has ordered a variation he has to act fairly in pricing it.

Pacific Associates Inc. v *Baxter* (1988) 44 BLR 33 (CA)

Pacific Associates contracted on the FIDIC form of contract to carry out dredging and reclamation works for the Ruler of Dubai. The defendant was employed as engineer to administer the contract. Disputes arose between the employer and contractor which went to arbitration and were subsequently settled with a £10 million payment to the contractor, with both parties paying their own costs. The contractor then brought a claim against the engineer for negligent certification, claiming the unrecovered balance of the claim together with interest and arbitration costs. His Honour Judge John Davies dismissed the claim. The contract contained a particular condition which stated:

> Neither any member of the Employer's staff nor the Engineer nor any of his staff, nor the Engineer's Representative shall be in any way personally liable for the acts or obligations under the Contract, or answerable for any default or omission on the part of the Employer in the observance or performance of any of the acts, matters or things which are herein contained.

The judge stated that 'the clear intention of (this clause) ... was to relieve the engineer of all personal liability for his acts and obligations under the contract' (13 Con LR 80, at page 93). He also stated that he felt the question of liability always depended on the particular terms of the contract in question, and that 'the over-riding intention of the contract was to put the engineer beyond the reach of legal responsibility for his acts' (at page 92). The contractor appealed but the appeal was dismissed. The Court of Appeal stated that the existence of the special clause meant that a duty of care could not in this case be imposed, but emphasised that otherwise such a duty might have existed.

7.3   Therefore, although not acting as adjudicator or in a quasi judicial role, the contract administrator must nevertheless act fairly and reach an independent decision. This obligation would normally be implied into SBC16, and failure to act in this way would be considered a breach of contract by the employer. Under-certification might seem an attractive option to the employer; however, at best it would offer only a short-term advantage. The contractor would be able to challenge the certificate in adjudication where, if successful, it would recover not only the shortfall, but also interest, and the employer would be required to pay the adjudicator's fees.

7.4   The payment provisions in SBC16 have been drafted to ensure compliance with the Housing Grants, Construction and Regeneration Act (HGCRA) 1996 Part II, as amended by the Local Democracy, Economic Development and Construction Act (LDEDCA) 2009. In summary, interim payment will normally follow the issue of an interim certificate, with provisions to ensure that the contractor is paid even if no certificate is issued. The contractor may submit its view as to what an interim certificate should cover by an 'Interim Application', but the valuation is a matter for the contract administrator. If no certificate is issued and an application is made, the amount stated in the application will become the amount due. Alternatively, if no certificate is issued and no application is made, after the date for issue of the certificate has passed the contractor may issue a payment notice. The amount shown in that notice will become the amount payable.

Advance payment

7.5   SBC16 includes an optional provision for advance payment to the main contractor. An entry must be made in the contract particulars to show whether or not it is to apply (cl 4.7). If it is, the amount will be entered as either a fixed sum or as a percentage of the contract sum. The entry must also show when payment is to be made to the contractor, and when it is to be reimbursed to the employer. A bond may be required, in which case payment is not made until the contractor provides the bond (SBC16 includes a form of advance payment bond as Part 1 of Schedule 6).

7.6   There is always a risk in making an advance payment with respect to a construction contract, even when backed by a bond, and the procedure will inevitably involve extra expense to the employer. The employer should be quite clear as to what compensatory benefits, such as a reduction in the contract sum, will result before agreeing to any arrangement of this sort. In the end it is the employer's decision, but the contract administrator may need to explain the provisions and give initial advice.

Interim certificates – timing

7.7   The due date for each interim payment is seven days from the 'Interim Valuation Date' (cl 4.8). The contract particulars require the parties to enter the first interim valuation date and, if none is entered, this is one month after the date of possession. Subsequent interim valuation dates are on the same date each month, or the nearest day which is not a Saturday, Sunday or bank holiday. The contract administrator must issue an interim certificate within five days of each due date (cl 4.9.1), and payment must be made within 14 days of the due date (cl 4.11.1). The effect of this is that unless the certificate is actually issued on the due date, the employer will have a shorter period within which it must pay.

Procedure for ascertaining amounts due

7.8   The amount due to the contractor as interim payment is to be calculated as set out in clauses 4.14 and 4.15. There are two methods by which it can be assessed: either through acceptance of the figure stated in a contractor's application for payment, or through valuation by the quantity surveyor (see Figure 7.1). The latter has traditionally been the basis for assessing payments under JCT forms.

7.9   Clause 4.10.1 gives the contractor the right to submit its own assessment of the value of an interim certificate. The 'Payment Application' must be made not later than the relevant interim valuation date and should be sent directly to the quantity surveyor. It is then open to the quantity surveyor to agree or disagree with this valuation. The quantity surveyor is not required to identify the matters where any differences arise or inform the contractor;

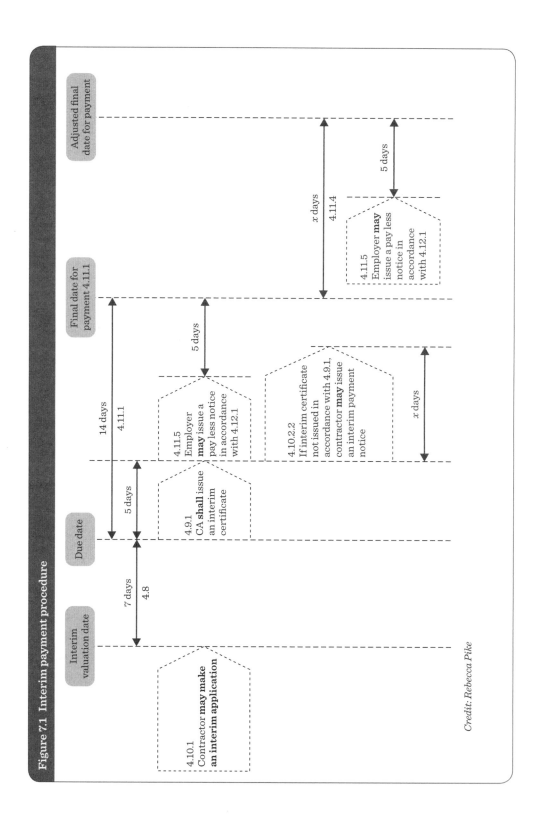

Figure 7.1 Interim payment procedure

Interim valuation date

Due date

Final date for payment 4.11.1

Adjusted final date for payment

7 days
4.8

5 days

14 days
4.11.1

5 days

5 days

x days
4.11.4

5 days

4.10.1
Contractor **may make an interim application**

4.9.1
CA **shall** issue an interim certificate

4.11.5
Employer **may** issue a pay less notice in accordance with 4.12.1

4.10.2.2
If interim certificate not issued in accordance with 4.9.1, contractor **may** issue an interim payment notice

x days

4.11.5
Employer **may** issue a pay less notice in accordance with 4.12.1

Credit: Rebecca Pike

however, in practice the quantity surveyor may respond to the application. The contract administrator should request that the quantity surveyor forwards copies of all correspondence and keeps the contract administrator informed regarding all applications.

7.10 Whether or not an interim application is received, the quantity surveyor will normally carry out a valuation prior to the due date (a valuation would be needed to determine whether the contractor's interim application is correct). The valuation should cover the amounts due at the interim valuation date (cl 4.14). The certificate is then issued 'not later than 5 days after each due date' (cl 4.9.1), which would allow it to be issued up to seven days before, on, or up to five days after the due date. Valuations by the quantity surveyor can be made 'whenever the Architect/Contract Administrator considers them necessary' (cl 4.9.2). The roles of the quantity surveyor and the contract administrator, however, are quite distinct, and it is ultimately for the contract administrator to decide the figure to be shown on the interim certificate.

Coverage of the certificate

7.11 Clause 4.9.1 requires that interim certificates state not only the sum that the contract administrator 'considers to be or have been due on the due date', but also 'the basis on which that sum has been calculated'. It is unlikely that a great deal of detail will be required here; a short schedule will probably be sufficient. Similar provisions are included for the final certificate.

7.12 The basis for calculating the sum due is set out in clause 4.15. This comprises the gross valuation of the work minus a list of deductions (see paragraph 7.15). The key items to be included in the gross valuation are set out in clause 4.14 and can be summarised as a total of 97 per cent (or as entered in the contract particulars) of the value of the following (i.e. items subject to retention):

- work properly executed, including variations, but excluding certain reinstatement work (cl 4.14.1.1);
- site materials (i.e. materials and goods properly on site) (cl 4.14.1.2);
- 'listed items' (cl 4.14.1.3).

7.13 The above amounts are to be adjusted in accordance with any fluctuations provisions and with any acceleration quotation that has been agreed (cl 4.14.1). The value of the work will be calculated using the rates shown in the bills or the priced document, whichever is appropriate. If a priced activity schedule is included, the amount included in any interim certificate in respect of any items listed in the activity schedule should be the total of the amounts reached by multiplying the percentage of the work properly executed by the price for that work as shown on the activity schedule (cl 4.14.1).

7.14 In addition to the above, the gross valuation should also include a total of 100 per cent of the following items, if applicable:

- costs associated with clause 2.6.2 (additional insurance premiums), clause 2.22 (royalties), clause 2.23 (patent rights), clause 3.17 (opening up and tests), clause 6.5 (optional insurance), clauses 6.10.2, 6.10.3 and 6.11.3 (terrorism cover), clause 6.12.2 (employer's insurance default) or clause 6.20 (Joint Fire Code) (cl 4.14.2.1);

- amounts payable under clause 4.13.2 (costs and expense reasonably incurred following suspension) (cl 4.14.2.2);

- loss and/or expense due under clause 4.20.1 or 5.3.3 (cl 4.14.2.3);

- sums for reinstatement work due under Insurance Option B or C, or under Option A to the extent that it is to be treated as a variation (cl 4.14.2.4);

- amounts payable under any applicable fluctuations provision that have not been adjusted under clause 4.14.1 (cl 4.14.2.5).

7.15   Before reaching the total gross valuation, the contract requires some deductions to be made if applicable:

- any amounts deductible under clause 2.10 (incorrect setting out), 2.38 and 3.18.2 (defects agreed not to be made good), clause 3.11 (costs incurred by the employer where instructions not complied with), clause 6.12.2 (contractor's insurance default) or clause 6.19.2 (Joint Fire Code) (cl 4.14.3.1);

- amounts allowable by the contractor in respect of clause 6.10.2 (terrorism cover) or any applicable fluctuations that have not been adjusted under clause 4.14.1 (cl 4.14.3.2).

7.16   Clause 4.14 states that the above assessment of the gross valuation is 'subject to any agreement between the Parties as to stage payments'. No other reference to stage payments is made in SBC16 (nor in the guide SBC/G). This is in contrast to DB16, where an alternative system using stage payments is set out in clause 4.12. Under a stage payment system, the contractor would not be paid anything for a stage until it is complete (unlike the 'activity schedule' system where the contractor would be paid a proportion even if the activity was not complete). If the parties would like to operate a stage payment system, they would need to agree in advance exactly how this would work, and include appropriate provisions in the contract – the procedure in DB16 could serve as a model.

Value of work properly executed

7.17   The contract administrator should only certify after having carried out an inspection to a reasonably diligent standard. Contract administrators should not include any work that appears not to have been properly executed, whether or not it is about to be remedied or the retention is adequate to cover remedial work (*Townsend* v *Stone Toms*, *Sutcliffe* v *Chippendale & Edmondson*). If an activity schedule is used, rather than a bill of quantities, this would not lessen the contract administrator's duty to determine that all work certified has been carried out in accordance with the contract. Contract administrators should also note the case of *Dhamija* v *Sunningdale Joineries Ltd*, which stated that a quantity surveyor is not responsible for determining the quality of work executed. If in doubt, the contract administrator may require 'reasonable proof' from the contractor that materials and goods comply (cl 2.3.4) and may carry out or request tests (cl 3.17). However, SBC16 makes it clear that interim certificates are not conclusive evidence that work is in accordance with the contract (cl 1.10).

7.18   Where work which has been included in a certificate subsequently proves to be defective, the value can be omitted from the next certificate. However, the contract does not appear to allow for a negative interim certificate. Clause 4.9.1 refers to the 'sum ... due to the Contractor' and does not include the provisions for payment to the employer that are set out for the final certificate (cl 4.26.2). Clause 4.12.2 also states that the amount 'may be zero' but does not refer to negative amounts. This clause derives from the LDEDCA 2009, and on balance it seems unlikely that the contract will be interpreted to allow for negative payments. If such a situation arises, and the employer does not wish to wait until the amount is recovered in subsequent certificates, it may be sensible to seek legal advice, particularly if the amount is significant.

Townsend v *Stone Toms & Partners* (1984) 27 BLR 26 (CA)

Mr Townsend engaged architect Stone Toms in connection with the renovation of a farmhouse in Somerset. John Laing Construction Ltd was employed to carry out the work on a JCT67 Fixed Fee Form of Prime Cost contract. Following the end of the defects liability period the architect issued an interim certificate that included the value of work which it had already included in its schedule of defects, and which it knew had not yet been put right. Mr Townsend brought proceedings against both Stone Toms and Laing. Laing made a payment into court of £30,000, which was accepted by Townsend in full and final settlement. Townsend then continued with the proceedings against the architect, claiming that he was entitled to recover any excess that he might have obtained from Laing had he continued with those proceedings. The Official Referee assessed the total value of the claims against Laing as only £25,000, therefore no excess was recoverable. The Deputy Official Referee also found that the architect was not negligent in issuing the interim certificate. Mr Townsend appealed and the Court of Appeal, although approving the lower court's decision on the effect of the payment into court, held that the architect had been negligent. Oliver LJ stated (at page 46):

> the whole purpose of the certification is to protect the client from paying to the builder more than the proper value of the work done, less proper retention, before it is due. If the architect deliberately over-certifies work which he knows has not been done properly, this seems to be a clear breach of his contractual duty, and whether certification is described as 'negligent' or 'deliberate' is immaterial.

Sutcliffe v *Chippendale & Edmondson* (1971) 18 BLR 149

(Note: this case is the first instance decision which was appealed to the Court of Appeal *sub nom. Sutcliffe* v *Thackrah*, discussed in paragraph 7.2 above.)

Mr Sutcliffe engaged the architect Chippendale & Edmondson in relation to a project to build a new house. No terms of engagement were agreed, but the architect proceeded to design the house, invite tenders and arrange for the appointment of a contractor on JCT63. Work progressed slowly and towards the end of the work it became obvious that much of the work was defective. The architect had issued ten interim certificates before Mr Sutcliffe entirely lost confidence, dismissed the architect and threw the contractor off the site. He then had the work completed by another contractor and other consultants, which cost around £7,000, in addition to which he was obliged, as a result of the original contractor having obtained judgment against him, to pay all ten certificates in full. As this contractor was subsequently declared bankrupt, Mr Sutcliffe brought a claim against the architect. The architect contended, among other things, that its duty of supervision did not extend to informing the quantity surveyor of defective work that should be excluded from the valuation. His Honour Judge Stabb QC found for Mr Sutcliffe, stating 'I do not expect that the words "work properly executed" can include work not then properly executed but which it is expected, however confidently, the Contractor will remedy in due course' (at page 166).

> *Dhamija* v *Sunningdale Joineries Ltd and others* [2010] EWHC 2396 (TCC)
>
> The claimants brought an action against the building contractor, the architect and the quantity surveyor (McBains) arising out of alleged defects in the design and construction of their home. There had been no written or oral contract with the quantity surveyor, but the claimants argued that there was an implied term that the quantity surveyor would only value work that had been properly executed by the contractor and was not obviously defective. The court held that a quantity surveyor's terms of engagement would include an implied term that the quantity surveyor act with the reasonable skill and care of a quantity surveyor of ordinary competence and experience when valuing the works properly executed for the purposes of interim certificates. However, the judge held that the quantity surveyor would not owe an implied duty to exclude the value of defective works from valuations, however obvious the defects. This was the exclusive responsibility of the architect appointed under the contract. Further, the quantity surveyor owed no implied duty to report the existence of defects to the architect.

Unfixed materials

7.19   Under the SBC16 provisions the contract administrator is obliged to include in the interim certificate materials which have been delivered to the site but are not yet incorporated in the works (cl 4.14.1.2), even though a limited risk to the employer remains (see paragraph 5.59). Clause 4.14.1.2, however, states that the obligation does not extend to materials that are prematurely delivered or not properly protected. Contract administrators should pay careful attention to the exact wording of this clause.

'Listed Items'

7.20   Interim certificates might include amounts in respect of off-site 'Listed Items' (cl 4.14.1.3). These may be 'materials, goods and/or items prefabricated for inclusion in the Works' and the items must be listed by the employer, and the list attached to the bills of quantities (or specification/schedules of work) and supplied to the contractor at tender stage (cl 1.1). The value of a listed item may be included in an interim certificate prior to its delivery on site provided certain preconditions are fulfilled:

- the listed item is in accordance with the contract (cl 4.16.1);
- the contractor has provided reasonable proof that the property is vested in it (cl 4.16.2.1);
- the contractor provides proof that the item is insured and will remain insured against specified perils until delivery on site (cl 4.16.2.2);
- the listed item is clearly identified as under order for the employer, and for the works, by being 'set apart' or clearly marked (cl 4.16.3);
- if the item is 'uniquely identified', a bond has been provided if required in the contract particulars (cl 4.16.4 and Schedule 6, Part 2);
- if the item is 'not uniquely identified', a bond has been provided (cl 4.16.5 and Schedule 6, Part 2).

7.21   It appears that the contract administrator has no discretionary power to certify any off-site items, other than those that have been listed. This makes the position for both parties clear, in that only 'listed' off-site materials are to be certified. Once certified, these items become the property of the employer, even though they are off-site (cl 2.25). The contract administrator should therefore be careful not to include any unlisted off-site materials in any certificate, as only listed items are covered by these protection mechanisms.

Costs and expenses due to suspension

7.22   Clause 4.13.2 states that 'Where the Contractor exercises his right of suspension under clause 4.13.1, he shall be entitled to a reasonable amount in respect of costs and expenses reasonably incurred by him as a result of exercising the right'. This amount is also to be included in interim valuations (cl 4.14.2.2). The phrase 'costs and expenses' is taken from the LDEDCA 2009, and suggests something more limited than the range of losses that could be claimed under a clause 4.20 loss and/or expense claim; for example, that it is limited to direct and ascertainable costs. However, a valid suspension would be due to a breach of contract by the employer, for which the contractor would be able to claim damages at common law, so it may be sensible for contract administrators not to interpret this too strictly.

Sum due

7.23   Clause 4.15 sets out the calculation that must be followed to reach the sum that is to be shown as due on each interim certificate. After calculation of the gross valuation, the following items are subtracted:

- any retention amounts that may be deducted;
- the total of any advance payments that are to be reimbursed;
- the sums stated as due in previous interim certificates;
- any amount paid in respect of a payment notice issued since the last certificate (this might occur if a previous certificate was not issued, and the contractor issued a payment notice in default, see paragraph 7.32).

Retention

7.24   Some of the items that must be included in the gross valuation are subject to retention (cl 4.14.1), of which half is released upon practical completion. As an alternative, the contractor may be required to provide a retention bond (cl 4.18). If this is to be used, the contract particulars must state that clause 4.18 is to apply, and the details must be set out at tender stage. SBC16 includes a form of retention bond as Part 3 of Schedule 6 (see paragraph 7.27). If a retention bond is not used, retention is deducted from the amounts set out in clause 4.14.1 (work properly executed, site materials and 'Listed Items').

7.25   The retention percentage is the amount inserted in the contract particulars, or if no amount is inserted is 3 per cent. The employer is trustee for the beneficiaries of the retention, i.e. the contractor (cl 4.17.1), and, except where the employer is a local or public authority, may be required under clause 4.17.3 to place the retention in a separate bank account. The contract administrator must issue a statement of the amount withheld with each interim certificate (cl 4.17.2), and the employer must certify that the amount has been placed in a separate account (cl 4.17.3). Clause 4.17.3 allows the employer the benefit of any interest which accrues. The employer has the right to deduct from the retention any sums due from the contractor (cl 4.12.3).

7.26   Retention has frequently been a point of controversy in the past. A series of cases established that the employer is obliged to place the money in a separate account, even if the contract contains no express provision, provided that it requires the employer to hold the money as a trustee (see *Wates Construction* v *Franthom Property* and *Finnegan* v *Ford Sellar Morris*). However, the court will not make an order to place money in a separate account following the insolvency of the employer (see *Mac-Jordan Construction* v *Brookmount Erostin*). The contractor would have no special claim beyond that of an unsecured creditor. To be safe the contractor must insist, while the employer is solvent, that the money is placed in a separate account.

Wates Construction (London) Ltd v *Franthom Property Ltd* (1991) 53 BLR 23 (CA)

Wates entered into a contract with Franthom on JCT80 to construct a hotel in Kent. Clause 30.5.3 (requiring Franthom to place retention in a separate account) had been deleted, but otherwise the retention clauses were in all material respects the same as those in SBC16. Although requested to do so by Wates, Franthom refused to place the accrued retention of around £84,000 in a separate account. Wates then commenced legal proceedings. Judge Newey ordered Franthom to place the money in an account, and Franthom then appealed. The court dismissed the appeal stating that 'clear express provisions are needed if a separate bank account is not to be set up'. The fact that the clause had been deleted did not of itself indicate what the parties' intentions were; the effect was the same as if the words had never been there at all.

J F Finnegan Ltd v *Ford Sellar Morris Developments Ltd* (1991) 53 BLR 38

Finnegan was the contractor on a JCT81 contract for works at Ashford. After the works reached practical completion the employer claimed liquidated damages of around £60,000 against a sum admitted as due to the contractor of around £20,000. Under clause 30.4.2.2 the employer was obliged to place retention monies deducted in a separate account if requested by the contractor. Finnegan commenced action to recover the sum due. The employer counterclaimed for the liquidated damages and Finnegan then requested that the retention be placed in a separate account. The employer refused and Finnegan applied for an injunction. The judge granted the injunction, despite the fact that this was long after practical completion. The contract did not require that a request was made each time retention was deducted nor at the time it was deducted.

Mac-Jordan Construction Ltd v *Brookmount Erostin Ltd* (1991) 56 BLR 1 (CA)

A developer held over £100,000 for the contractor in retention money but was also heavily indebted to the bank (floating loan granted by a charge). The developer went into insolvency and the bank appointed administrative receivers. The contractor then sought a court injunction to establish a separate retention fund, but the Court of Appeal refused on grounds that this would give an unsecured creditor (the contractor) preference over any other unsecured creditors of an insolvent debtor. The contractor's right to the retention was stated to be no more than an 'unsatisfied and unsecured contractual right for the payment of money' (Scott LJ at page 15).

Bond in lieu of retention

7.27 This dilemma over retention and the effectiveness of trustee status has raised the question of bonds and guarantee bonds from both employer and contractor, respectively, as an alternative, and SBC16 makes provision for the use of a bond in lieu of retention (cl 4.18). If a bond is to be required, then the contract particulars must indicate that clause 4.18 is to apply and specify the maximum aggregate sum to be secured. The form of bond to be used is included in Schedule 6, Part 3, and must be provided by the contractor prior to the date of possession.

7.28 Where clause 4.18 applies, retention is not deducted from amounts on certificates. Instead, a statement of the retention that would have been deducted is prepared prior to each interim certificate. If at any time this statement exceeds the maximum aggregate sum stated in the bond, either the contractor arranges for the bond to be adjusted or the employer deducts retention for the unsecured amount. If the contractor fails to provide the bond at all, the employer may deduct retention as described above.

Advance payments and bonds

7.29 The advance payment indicated in the contract particulars is paid to the contractor before the first certificate of payment is due for issue, but only after the contractor has provided the bond required (cl 4.7). Payment is made directly from the employer to the contractor, and the contract administrator should ensure that it receives copies of any correspondence regarding this. Details of when the reimbursements are to take place will also be set out in the contract particulars and could, for example, be in stages throughout the project. The reimbursement is deducted from the gross valuation under the relevant certificate (cl 4.15.2). It is not clear why reimbursement of advance payments is to be shown on certificates, as the original payment would not appear.

VAT

7.30 The contract sum is exclusive of any VAT (cl 4.5.1), and there is no requirement to indicate the VAT applicable to any certified amount. VAT is not a matter of contract but of statute and is normally paid by the employer on submission of a VAT invoice by the contractor following each interim certificate.

Payment procedure

7.31 The final date for payment of each interim certificate is 14 days from the due date (cl 4.11.1) (see Figure 7.1). If the employer intends to withhold any amount from the sum certified, then the contractor must be given written notice of this no later than five days before the final date for payment, in the form of a 'Pay Less Notice' (cl 4.11.5). The pay less notice should be issued by the employer, unless the employer has notified the contractor that the contract administrator, quantity surveyor or other person is authorised to issue the notice on its behalf (cl 4.12.1.1). The notice should set the sum that the employer considers is due to the contractor at the date the notice is given and 'the basis on which that sum has been calculated' (cl 4.12.1). If a pay less notice is issued, the employer must pay the contractor at least the sum set out in that notice by the final date for payment (cl 4.11.5).

Payment when no certificate is issued

7.32   The contract has a remedy in the event that the date for issue has passed and no certificate has been provided. If the contractor has not already submitted a payment application, it can at any time after the issue date has passed send a payment notice to the quantity surveyor (cl 4.10.2.2). The notice should state 'the sum that the Contractor considers to have been due to him under clauses 4.14 and 4.15 or clause 4.26.2 at the relevant due date and the basis on which that sum has been calculated'. The sum shown on the notice will become the amount due. If the contractor has already submitted a payment application, it need take no further action, as that application will be considered a payment notice (cl 4.10.2.1). Where payment notice is given, clause 4.11.4 states that 'the final date for payment of the sum specified in it shall for all purposes be regarded as postponed by the same number of days as the number of days after the last date for issue of the Payment Certificate that the Payment Notice is given'. This is best understood by way of example; if the payment notice was issued six days after the final date for issue of the certificate (i.e. 11 days in total after the due date), the final date for payment would be 20 days after the due date (i.e. 14 plus six days). If the employer disagrees with the amount shown on the payment notice, then it must issue a pay less notice as described above (which in this example would be within four days of the notice). It should be noted that where no certificate is issued, the employer may not issue a pay less notice until after the contractor has submitted a payment notice (cl 4.13.1.1; i.e. it cannot use the pay less notice as a substitute for the missing certificate).

Deductions

7.33   As noted above, the contract administrator is required to make certain deductions before reaching the gross valuation. These are listed in clause 4.14.3, and include amounts relating to the acceptance of contractor errors (cl 2.10, 2.38 and 3.18.2), costs due to failure to comply with instructions (cl 3.11), and some insurance costs (cl 6.12.2) and fluctuation sums. In addition, in calculating the 'sum due' on a certificate, the contract administrator must make further deductions as listed in clause 4.15 (see paragraph 7.23). The employer then has the right to issue a pay less notice, which may state a different sum due. The pay less notice could, for example, make any of these deductions, if not already accounted for, or correct some arithmetical error in the certificate. In addition, the employer may make a deduction of liquidated damages (cl 2.32.2, see paragraph 4.62).

7.34   In addition, the employer in some circumstances would have the right to make deductions for defective work. Prior to the HGCRA 1996, it was clear that if the employer had an arguable case that the certificate may have included work which was defective, and therefore had been overvalued, then the employer need not have paid the full amount, but could have raised the losses due to the defects either as a counterclaim in any action brought by the contractor or as a defence to the claim. The latter process is often termed 'abatement' by lawyers.

7.35   It is now generally agreed that in cases where such a right may exist, it can only be exercised through the use of the 'Pay Less Notice' procedure, as discussed above. The employer would therefore be unable to withhold amounts to cover any defective work included in a certificate, unless the deduction is covered by a notice (*Rupert Morgan Building Services* v *Jervis*). In fact the courts have consistently upheld the contractor's right to be paid the full amount certified (or notified) in the absence of a valid pay less notice, despite numerous and often ingenious attempts by employers to avoid payment (see for example *Kersfield Developments* v *Bray and Slaughter*).

Rupert Morgan Building Services (LLC) Ltd v David Jervis and Harriet Jervis [2004] BLR 18 (CA)

A couple engaged a builder to carry out work on their cottage, by means of a contract on the standard form published by the Architecture and Surveying Institute (ASI). The seventh interim certificate was for a sum of around £44,000 plus VAT. The clients accepted that part of that amount was payable but disputed the balance, amounting to some £27,000. The builder sought summary judgment for the balance. The clients did not give 'a notice of intention to withhold payment' before 'the prescribed period before the final date for payment'. The builder contended that it followed, by virtue of HGCRA 1996 section 111(1), that the clients 'may not withhold payment'. The clients maintained that it was open to them, by way of defence, to prove that the items of work which go to make up the unpaid balance were not done at all, or were duplications of items already paid or were charged as extras when they were within the original contract, or represented 'snagging' for works already done and paid for. The Court of Appeal determined that, in the absence of an effective withholding notice, the employer has no right of set-off against a contract administrator's certificate.

Kersfield Developments (Bridge Road) Ltd v Bray and Slaughter Ltd [2017] EWHC 15 (TCC)

Contractor Bray and Slaughter entered into a contract with Kersfield to refurbish a mansion house. The contract was on JCT DB11, and in the sum of £5 million. After Kersfield failed to pay Bray's payment application number 19, Bray obtained an adjudication decision in its favour for £1.2 million based on Kersfield's failure to serve a payment or pay less notice within the contractual time limits. Kersfield commenced TCC proceedings. It asked the court for a declaration that it was entitled to commence a further adjudication to determine the correct value of Bray's payment application, which it was now disputing. Kersfield relied on paragraph 20 of the Scheme adjudication rules, which states that an adjudicator may 'open up, revise and review any decision taken or any certificate given by any person referred to in the contract unless the contract states that the decision or certificate is final and conclusive'. The court rejected this argument, finding that payment and pay less notices did not qualify as 'decisions' or 'certificates' under paragraph 20 of the Scheme. In doing so, it followed the principles set out in earlier cases, such as *ISG Construction Ltd v Seevic College* [2014] EWHC 4007 (TCC) and *Galliford Try Building Ltd v Estura Ltd* [2015] EWHC 412 (TCC).

Contractor's position if the certificate is not paid

7.36   SBC16 includes several provisions which protect the contractor if the employer fails to pay the contractor amounts due. Clause 4.11.6 makes provision for simple interest to accrue on any unpaid amount. The defined 'Interest Rate' is set at 5 per cent above the official bank rate of the Bank of England, unless the contract particulars state some other rate, and the interest accrues from the final date for payment until the amount is paid. Similar provisions are included for the final certificate and for amounts due under sub-contracts. If the employer makes a valid deduction by means of a pay less notice, it is suggested that interest would not be due on this amount. The clause does not refer to the amount stated on the certificate but to 'a sum, or any part of it' due to the contractor under the conditions, which would take into account valid deductions.

7.37   The contractor is also given a 'right of suspension' under clause 4.13. This right is required by the HGCRA 1996 Part II. If the employer fails to pay the contractor by the final date for payment, the contractor has a right to suspend performance of all its obligations under the contract, which would include not only the carrying out of the work, but could also, for example, extend to any insurance obligations. This non-payment is stated to be of 'the sum payable in accordance with clause 4.11'; therefore, the contractor may not suspend work if a pay less notice has been issued by the employer. The contractor must have given the employer written notice of its intention to suspend work and stated the grounds for the suspension, and the default must have continued for a further seven days. The contractor must resume work when the payment is made. Under these circumstances the suspension would not give the employer the right to terminate the contractor's employment. Any delay caused by the suspension is a relevant event (cl 2.29.6) and any reasonable costs and expenses incurred are claimable under clause 4.13.2 (see paragraph 7.22).

7.38   The contractor has the right to terminate its employment under the contract if the employer does not pay amounts due (cl 8.9.1.1). The contractor must give notice of this intention, which specifies the default as required by the contract.

Contractor's position if it disagrees with amount certified

7.39   As a result of the introduction of the payment notice provisions, it is no longer the case that the issue of a certificate is a condition precedent to the right of the contractor to be paid. However, where a certificate has been issued, the contractor is only entitled to the amount shown, even if the certificate is undervalued and irrespective of what the contractor may have put forward in its interim application. This case of *Lubenham v South Pembrokeshire District Council* states that the contractor is only entitled to the sum stated in the certificate, even if the certificate contains an error, for example because it includes a wrongful deduction. The contractor's remedy is to request that the error is corrected in the next certificate, or to bring proceedings, for example adjudication, to have the certificate adjusted. (There are exceptions to this rule, for example where the employer has interfered with the issue of the certificate, in which case the contractor may be entitled to summary judgment for the correct amount.) This appears to be the case even where the interim certificate shows a negative amount, i.e. an amount payable by the contractor to the employer, which may occur if the contract administrator decides that work was overvalued or there was some other error in an earlier certificate. Unlike the final certificate, the clauses relating to interim payments do not give the contractor the right to issue a pay less notice.

> *Lubenham Fidelities and Investments Co. Ltd v South Pembrokeshire District Council* (1986) 33 BLR 39 (CA)
>
> Lubenham Fidelities was a bondsman which elected to complete two building contracts based on JCT63. The architect, Wigley Fox Partnership, issued several interim certificates which stated the total value of work carried out, but also made deductions for liquidated damages and defective work from the face of the certificate. Lubenham protested that the certificates had not been correctly calculated, withdrew its contractors from the site and issued notices to determine the contract. Shortly afterwards, the Council gave notice of determination of the contract. Lubenham brought a claim against the Council on the grounds that its notices were valid and effective, and against Wigley Fox on the basis that the architect's negligence had caused it losses. It was held that the Council was not obliged to pay more than the amount on the certificate and that, whatever the cause of the undervaluation, the correct procedure was not to withdraw labour but to request that the error was corrected in the next certificate, or to pursue the matter in arbitration. Lubenham's claim against Wigley Fox failed because it had been the suspension of the works rather than the certificates that had caused the losses, and because the architect had not acted with the intention of interfering with the performance of the contract.

Interim payment on practical completion

7.40   The effect of clause 4.19.2 is to release to the contractor half of the retention that has been deducted in the first interim certificate following practical completion of the works, or in the case of a section, half the proportion that relates to that section. Following practical completion, certificates continue at monthly intervals (cl 4.8), and the employer retains the right to deduct half the retention percentage from the amounts due under these certificates. Although no further work should be carried out during this period (except instructed remedial work), amounts may nevertheless become due; for example, when a claim for loss and/or expense has not been resolved prior to practical completion.

Final certificate

7.41   To summarise, by final certificate stage the following certificates should have been issued:

● interim certificates at monthly intervals, with the full retention percentage applied (cl 4.8, cl 4.19.1);

● practical completion certificate (cl 2.30);

● certificates at monthly intervals during the rectification period, with half the retention percentage applied (cl 4.8, cl 4.19.2);

● certificate of making good (cl 2.39).

7.42   The final certificate must be issued within the specific time periods set out in the contract (cl 4.26.1) (see Figure 7.2). In practice, the latest date for issue tends to be determined by the process of calculating the adjusted contract sum. The onus is on the contractor to send all necessary information to the contract administrator or, if instructed to do so, to the quantity surveyor, not later than six months after practical completion of the works (cl 4.25.1). No later than three months after receiving this information, the contract administrator, or the quantity surveyor if asked to do so, makes a final assessment of the amount of loss and/or expense due (cl 4.25.2.1), and the quantity surveyor prepares a statement of all adjustments to be made to the contract sum (cl 4.25.2.2). This must be sent to the contractor 'within that 3 month period' (cl 4.25.2).

7.43   If the contractor fails to supply the necessary information within six months of practical completion, then the contract administrator may give a one-month notice requiring its supply (cl 4.25.2.3). If the contractor does not comply with the notice, the contract administrator and quantity surveyor may proceed on the basis of the information that they already have. Note that there is no requirement to issue a notice or proceed with the assessments, this is merely a right, so matters could come to a standstill if the contractor does not act. In some cases, however, the employer may prefer to finalise matters, despite the lack of action by the contractor.

7.44   The final certificate is then issued within two months of the statement (or, within two months of the issue of the last certificate of making good, or the expiry of the last rectification period, whichever is the latest) (cl 4.26.1). It is worth noting that it has been held that the final certificate can be issued at the same time as the statement, although it would be good practice to allow the contractor time to consider the documents (*Penwith District Council* v *V P Developments*). It is also worth noting that a document does not necessarily have to be headed 'Final Certificate' or be in any particular format to comply

Figure 7.2 Final payment procedure

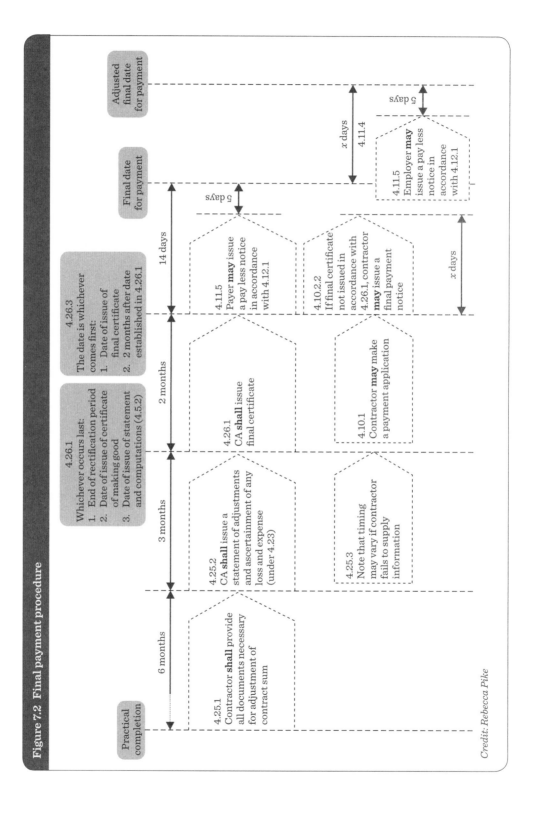

Credit: Rebecca Pike

> ### Penwith District Council v V P Developments Ltd [1999] EWHC Technology 231
>
> Penwith District Council employed VP for maintenance works to 91 houses at Hayle. The contract was on JCT80. Practical completion took place on 21 September 1990, and the certificate of making good defects was issued on 30 October 1991. VP submitted a draft final account on 14 January 1991. Three interim certificates were issued following practical completion, the last one on 10 July 1992. The final certificate was issued on 8 April 1993, and enclosed a document summarising how the figure on the final certificate had been arrived at. VP gave notice of arbitration some three years later. It argued that it was not barred by the clause 30.9 conclusiveness provisions as the final certificate had not been valid. The arbitrator found for VP, stating that the intention of the contract was that the contractor should have at least three months to consider the ascertainment of final account referred to in clause 30.6.1. The Council appealed and His Honour Judge Humphrey Lloyd found that the contract terms required that no minimum period should have elapsed, all the time limits referred to were maxima. He also found that no such term could be implied: 'the 1980 JCT form is a long and complex document and was plainly intended to provide for most conceivable circumstances and to block the many attempts to find gaps in its structures, despite repeated assaults'.

> ### B R Cantrell (2) E P Cantrell v Wright and Fuller Ltd [2003] BLR 412
>
> Cantrell engaged Wright on a JCT80 form to construct an extension to a nursing home in Woodbridge, Suffolk. The work achieved practical completion on 23 February 1998. Following practical completion the contract administrator did not issue an extension of time, nor had it issued a certificate of non-completion. Following meetings between quantity surveyors, a document entitled 'final account' was agreed. On 12 March 1999 the contract administrator sent the claimant the final account plus an interim certificate in the sum of around £25,000. On 29 March 1999 the contract administrator issued a further certificate, which referred to a 'final payment', and was accompanied by a letter which referred to it as a 'final certificate'. The employer's solicitors immediately challenged the adjustments to the contract sum, and the contractor's solicitors demanded payment. In May 2002 a notice of arbitration was served and an arbitrator appointed. The parties were in dispute as to whether the March 1999 document was a final certificate complying with clause 30.8. If it was, then its conclusive effect would defeat some of the matters claimed. The arbitrator decided that the certificate was a final certificate.

with clause 4.26: whether it constitutes a final certificate will depend on the facts in each case (*Cantrell* v *Wright and Fuller*).

7.45 The final certificate must state the contract sum as adjusted under clause 4.3, which sets out all the deductions and additions to the contract sum (cl 4.26.2.1). The final date for payment of the final certificate is 14 days from the due date, which is the date of issue of the final certificate, or the last date of the period in which it should have been issued, whichever is the earlier (cl 4.11.1 and cl 4.26.3). The contract acknowledges that the final certificate can be for a negative amount – in other words, it can certify that payment is due from the contractor to the employer (cl 4.26.2). The final certificate is subject to pay less notice provisions, as described above in relation to interim certificates, except that in this case the contractor is also given the right to issue a pay less notice (cl 4.11.5.2).

Payment when no final certificate is issued

7.46 If no final certificate is issued within the required two-month period then, as with interim payments, the contractor may then send a final payment notice to the employer (cl 4.10.2), in which case the final date for payment would be postponed (see paragraph 7.32). The notice should state the sum that the contractor considers to be due and the basis on which that sum has been calculated (cl 4.10.1). If the employer disagrees with the amount shown on the notice then it may issue a pay less notice under clause 4.11.5. The employer is then only obliged to pay the amount shown in the pay less notice.

7.47   In the unlikely event that neither a final certificate nor a final payment notice is issued, the contract nevertheless provides that there is a date by which a final payment should be made. Clause 4.26.3 states that 'The due date for the final payment shall be the date of issue of the Final Certificate or, if that certificate is not issued within the 2 month period referred to in clause 4.26.1, the last day of that period'. This provides protection to the contractor, in that even if it does not issue a final payment notice, the contract establishes a date by which final payment must be made, i.e. 14 days from the due date. The employer would therefore be wise to make a payment, regardless of the lack of certificate or payment notice. Interest will start to run from that date, as set out in clause 4.11.6, based on the sum which is later determined should have been paid at that time. This would be the case even if neither the contract administrator nor the contractor have finalised their calculations regarding the final sums due.

Conclusive effect of final certificate

7.48   The final certificate is conclusive evidence that proper adjustment has been made to the contract sum (cl 1.9.1.2), and the contractor is prevented from seeking to raise any further claims for extensions of time (cl 1.9.1.3) or for reimbursement of direct loss and/or expense (cl 1.9.1.4). It is also conclusive evidence that where matters have been expressly stated to be for the approval of the contract administrator they have been approved (cl 1.9.1.1), but apart from those matters it is not conclusive that any other materials, workmanship, etc. are in accordance with the contract.

7.49   If dispute resolution proceedings are commenced before the issue of the final certificate, or within 28 days of the date of its issue, the conclusive effect is 'suspended' (the certificate does not become conclusive) in relation to the matters the subject of dispute until the proceedings are concluded (cl 1.9.2). Proceedings are concluded when a decision, award or judgment is issued, or if either party takes no further action for a period of 12 months (cl 1.9.3). In cases where a dispute has been raised in adjudication and the adjudicator's decision is reached after the final certificate, if either party wishes to challenge that decision then it must initiate proceedings within 28 days of the date of the decision (cl 1.9.2.2). The bar on raising matters after the 28-day period cannot be extended by the court, as the bar is an evidential bar and not a bar to bringing proceedings. In other words, an arbitration could be commenced, but no evidence can be brought forward.

7.50   The conclusive effect of the final certificate was the subject of much heated debate following the decisions in *Colbart Ltd* v *H Kumar* (1992) 59 BLR 89 and *Crown Estate Commissioners* v *John Mowlem*, cases which would still apply to older versions of JCT forms (see *London Borough of Barking & Dagenham* v *Terrapin*). Following the cases, the JCT amended the relevant clauses in its contracts. It appears unlikely, under the current wording, that an employer would be unable to effectively raise a claim regarding work or materials which were not in accordance with the contract following the 28-day cut-off period, provided that it had not been stated to be 'to approval' of the contract administrator somewhere within the contract documents.

Crown Estate Commissioners v *John Mowlem & Co. Ltd* (1994) 70 BLR 1 (CA)

Crown Estates employed Mowlem to construct a commercial development on the site of the former Kensington Palace Barracks. A final certificate was issued on 2 December 1992, and on 6 April 1993, Crown Estates gave notice of arbitration. It then issued a summons under section 27 of the Arbitration Act 1979 for an order extending the time within which to commence arbitration, in order to validate its notice. In addition to the summons the judge at first instance was also asked to consider the question of what, if anything, the final certificate was conclusive evidence of, as this would affect what could be raised in the arbitration. The judge issued the order extending time and held that the final certificate was only conclusive as to matters that were expressly stated to be for the satisfaction of the contract administrator. Mowlem appealed and the appeal was allowed. The Court of Appeal stated that clauses 30.9.1.1 and 30.9.3 did not limit the time within which arbitration proceedings could be brought, therefore the court had no powers under the Arbitration Act 1979 that could defeat the effect of the certificate. It also held that as all standards and quality of work and materials were inherently matters for the opinion of the contract administrator, the final certificate was conclusive evidence of all such matters.

London Borough of Barking & Dagenham v *Terrapin Construction Ltd* [2000] BLR 479

The Borough employed Terrapin Construction to design and build new and refurbishment work at a school in Dagenham. No document entitled 'Employer's Requirements' had been issued to the contractor at tender stage, but the contractor had been given a 'brief', which set out in general terms the nature of works which the Borough wanted to have designed, and the court decided that requirements were 'represented by the contract as a whole'. The contract was to be on WCD81. Once a tender figure had been negotiated, the Borough sent the contractor an order for the work, which set out the agreed contract figure, incorporated the terms of WCD81 and also stated: 'In consideration of this Agreement hereinafter contained on the part of the employer the Contractor shall and will execute complete and maintain the Works in all respects to the satisfaction of the Controller of Development and Technical Services'. The court decided that in this context the final statement was conclusive evidence that all work had been carried out to the satisfaction of the employer.

7.51   In cases where work has been stated to be 'to approval', the contract administrator should take particular care and should not issue the certificate unless and until the work is satisfactory. Where unsatisfactory work has been accepted, the contract administrator should, together with the employer, weigh the advantages of issuing the final certificate (the element of finality it brings to matters such as the final account) against the disadvantages (not being able to claim with respect to the unsatisfactory work) before deciding whether to proceed.

8 Insurance

8.1 Construction operations can be hazardous, and it is therefore important that liability for losses resulting from personal injury or damage to any property or to the works is clearly allocated to one party or the other, and that the liability is backed up by insurance. Should any incidents occur, it is vital that there should be no room for dispute about who is liable for the losses, and that all concerned should be clear about what procedural steps must be taken. Ambiguity in the contract can only lead to confusion and delays, which will benefit neither party.

8.2 It is usual for the contractor to indemnify the employer in respect of certain losses – for example for injury to persons, or for damage to neighbouring property which has been caused by the contractor's negligence – and in SBC16 this is done under clauses 6.1 and 6.2. This indemnity protects the employer in that if an injured party brings an action against the employer, rather than against the contractor, the latter has agreed to carry the consequences of the claim. If a third party sues the employer, then the employer can join the contractor as co-defendant or bring separate proceedings. Indemnities given to the employer by the contractor will obviously be quite worthless unless there are adequate resources to meet claims. The contract therefore requires insurance cover under clause 6.4 to back up the indemnities.

8.3 In addition to the requirement for insurance against claims arising in respect of persons and property, the contract contains alternative provisions for insurance of the works. There is also an optional provision (cl 6.5.1) for insurance against damage caused to property which is not the result of the negligence of the contractor, if required by the employer.

Injury to persons and damage to property

8.4 Clauses 6.1–6.3 cover injury to persons and damage to property other than the works, which arise from the carrying out of the works. The contractor is required to match the indemnities given in clauses 6.1 and 6.2 with insurance under clause 6.4 (cl 6.4.1). The contractor must be able to provide evidence that this insurance has been taken out (cl 6.4.2 and 6.12). The minimum cover required as a contractual obligation is entered in the contract particulars. If the contractor defaults, the employer may take out the insurance and deduct the cost from any sums due or to become due to the contractor, or recover them as a debt (cl 6.12.2).

8.5 Clause 6.4.1.1 requires that the insurance in respect of personal injury or death of any person in a contract of service with the contractor should comply with 'all relevant legislation'. The contractor's liability in respect of personal injury or death of employees is met by an employer's liability policy. This has been compulsory since the Employers' Liability (Compulsory Insurance) Act 1969. The statutory requirement is for a cover level of £5 million, although in practice most standard policies provide cover of at least £10 million.

8.6   The contractor's liability in respect of third parties (death or personal injury and loss or damage to property, including consequential loss) is met by its public liability policy. Insurers advocate insuring for a minimum of £2 million for any one occurrence, although a higher amount may be required by some clients. The contractor is required to insure the indemnities required under clause 6 up to the amount stated in the contract particulars (cl 6.4.1.2). Liability at common law for claims by third parties is unlimited, and any amount specified in the contract is merely the employer's requirement in the interests of safeguarding against inadequacies, and in no way limits the contractor's liability under clauses 6.1 and 6.2. It is recognised in footnote [50] to clause 6.4.1.2 that it may not always be possible to acquire insurance cover which is co-extensive with the indemnity required in clauses 6.1 and 6.2. For example, the insurance market has removed gradual pollution from its public liability policies. This again does not affect the contractor's duty to indemnify.

8.7   The liability and duty to indemnify are subject to exceptions. In respect of liability for personal injury or death, this is qualified in that the contractor is not liable where injury or death is caused by an act of the employer, or a person for whom the employer is responsible (cl 6.1).

8.8   In respect of damage to property, the contractor is only liable to the extent that the damage is caused by negligence or breach of statutory duty or other default of 'the Contractor or any Contractor's Person' (cl 6.2). The contractor is therefore liable only for losses caused by its own defaults. It is made clear in clause 6.3.4 that the definition of 'property' excludes the works, up to practical completion of the works or a section, except for parts taken over by partial possession. The contractor is therefore liable for any damage it negligently causes to the property of third parties, or to parts of the works taken over by partial possession (cl 6.3.4.2), or to a section of the works, after the relevant section completion certificate is issued (cl 6.3.4.1) or to the works after practical completion.

8.9   Clause 6.3.1 also excludes, where Insurance Option C applies, liability for 'any loss or damage to Existing Structures or to any of their contents required to be insured under that option that is caused by any of the risks or perils required or agreed to be insured against under that option'. This means that, where Insurance Option C is applicable, the contractor is not liable for losses insured under paragraph C.1 and caused by the listed perils, even where the damage is caused by the contractor's own negligence. This point is now expressly stated in clause 6.3.2 (where a paragraph C.1 replacement schedule is used, the liability is subject to exclusions and limitations set out under that schedule; cl 6.3.3, see paragraph 8.19). The exclusion was inserted to clarify matters following a series of cases on older versions of JCT contracts that reached the opposite conclusion (*National Trust* v *Haden Young*, *London Borough of Barking & Dagenham* v *Stamford Asphalt Co.*). It should be noted that the contractor might remain liable for some consequential losses (*Kruger Tissue* v *Frank Galliers*).

> *The National Trust for Places of Historic Interest and Natural Beauty v Haden Young Ltd* (1994) 72 BLR 1 (CA)
>
> The National Trust employed a contractor to carry out repair works to Uppark House, South Harting, West Sussex. The main contract was on terms substantially similar to MW80. Haden Young was sub-contractor for the renewal of lead work on the roof. During the course of the works a fire broke out, causing extensive damage, which Haden Young admitted was caused

by the negligence of its workforce, and the National Trust brought a claim for damages. Otton J found the sub-contractor liable at first instance, and that the employer's liability to insure under clause 5.4B only extended to matters not caused by negligence. Clauses 5.2 and 5.4B formed a coherent and mutually supportive structure. Haden Young appealed, but the appeal was dismissed. Although the Court of Appeal agreed that the sub-contractor was liable, it disagreed with the reasoning of the lower court, stating that there was no reason why there should not be an overlap, in other words why the employer should not be required to insure for matters for which the contractor was liable under clause 6.2. However, the damages recoverable from the contractor under clause 6.2 would be reduced by the amount recoverable by the employer under the clause 6.3B insurance.

London Borough of Barking & Dagenham v *Stamford Asphalt Co. Ltd* (1997) 82 BLR 25 (CA)

Barking & Dagenham employed a contractor to carry out repair works to a school. The main contract was on MW80, 1988 revision. Stamford was sub-contractor for the renewal of lead work on the roof. During the course of the works a fire broke out, causing extensive damage, which Stamford admitted was caused by the negligence of its workforce, and the Borough brought a claim for damages. The Court of Appeal found the contractor liable for the damage caused, preferring the reasoning of Otton J in *National Trust* v *Haden Young* to that of the Court of Appeal in that case. It should be noted that the wording of clause 6.3 has been adjusted to make it clear that the contractor is not liable for damage to property insured under paragraph C.1.

Kruger Tissue (Industrial) Ltd v *Frank Galliers Ltd* (1998) 57 Con LR 1

Damage was caused to the existing building and works by fire, assumed for the purposes of the case to be the result of the negligence of the contractor or sub-contractor. The construction work being carried out was on a JCT80 form. The employer brought a claim for loss of profits, increased cost of working and consultants' fees, all of which were consequential losses. Judge John Hicks decided that the employer's duty to insure for 'the full cost of reinstatement, repair or replacement' of the existing structure and the works under clause 22C (and therefore contractor's exemption from liability under clause 20.2) did not include such consequential losses. A claim could therefore be brought against the contractor for these losses. (Note that SBC16 now provides for professional fees coverage to be required as part of the works insurance.)

Damage to property not caused by the negligence of the contractor

8.10   The liability for damage to adjoining buildings where there has been no negligence on the part of the contractor is not covered under clause 6.2. Subsidence or vibration resulting from the carrying out of the works might cause such damage, even though the contractor has taken reasonable care. This is a risk which may be quite high in certain projects on tight urban sites, or in close proximity to old buildings. In such cases it may be advisable to take out a special policy for the benefit of the employer.

8.11   In SBC16 there is an optional provision for this type of insurance under clause 6.5.1. If it is anticipated that the main contractor may be required to take out this insurance, the correct deletion must be made in the contract particulars, and the amount of cover entered. The contract administrator must then instruct the contractor to 'effect and maintain' the policy, after confirming with the employer that the policy is required. The cost is added to the contract sum. The policy must be in joint names and placed with insurers approved by the employer. The policy and receipt are to be deposited with the contract administrator or, if so instructed, with the employer (cl 6.5.2).

8.12   This insurance is usually expensive, and subject to a great many exceptions. If it is required, then the policy needs to be effective at the start of the site operations when demolition, excavation, etc. are carried out. The text of clause 6.5.1 was revised in 1996 to take account of the wording of model exclusions compiled by the Association of British Insurers. The policy should be checked by the employer's insurance advisers to ensure that any exclusions correlate with clause 6.5.1 and that the policy provides the cover that the clause requires.

Insurance of the works

8.13   There are three alternative options for covering insurance of the works (Options A, B and C), which are set out in Schedule 3, and the option which is applicable should be entered in the contract particulars (cl 6.7). If either party defaults in taking out the required insurance, the other may take out the policy and the defaulting party will be liable for the costs (cl 6.12.2).

8.14   In all three alternatives, the policies are to be in joint names, and cover must be maintained up until practical completion of the works, or termination, if this should occur earlier (Schedule 3 and cl 6.7.2). The 'Joint Names Policy' definition was reworded in 1996 to make clear the intention that, under the policy, the insurer does not have a right of subrogation to recover any of the monies from either of the named parties. The policies must also either cover sub-contractors or include a waiver of any rights of subrogation against them (cl 6.9.1). This coverage is in respect of specified perils only, and not the full range of risks covered by 'All Risks Insurance'.

8.15   Insurance Options A and B are for insuring new building work and require 'All Risks' cover under joint names policies. A definition of 'All Risks' is given in clause 6.8 and refers to 'any physical loss or damage to work executed and Site Materials and against the reasonable cost of the removal and disposal of debris'. There is also a list of exclusions, which includes the cost necessary to repair, replace or rectify property which is defective, loss or damage due to defective design, loss or damage arising from war and hostilities and 'Excepted Risks' (except as provided by terrorism cover). Footnote [54] explains that cover should not be reduced beyond the exclusions set out in the definitions. It also points out that 'All Risks' cover that includes the risk of defective design, although not required, may be available. If the policy provided is likely to differ in any way from the requirements in the contract, this must be discussed and agreed before the contract is entered into. Even in a so-called 'All Risks' insurance policy there may be further exclusions, and the employer's insurance advisers should carefully check the wording of each policy.

8.16   Option A insurance is taken out by the contractor and is to be for the full reinstatement value of the works, including professional fees to the extent entered in the contract particulars (A.1). If no percentage is stated for professional fees, the default rate is

15 per cent. The contractor is responsible for keeping the works fully covered and, in the event of underinsurance, will be liable for any shortfall in recovery from the insurers.

8.17   Option B insurance is taken out by the employer, and again is to be for the full reinstatement value of the works, including professional fees (B.1). The employer is responsible for keeping the works fully covered and, in the event of underinsurance, will be liable for any shortfall.

8.18   Option C is applicable where work is being carried out to existing buildings. It includes two insurances, both taken out by the employer. The existing structure and contents must be insured against 'Specified Perils' as defined in clause 6.8 (C.1). (The main difference between 'All Risks' and 'Specified Perils' is the omission in the latter definition of risks connected with impact, subsidence, theft or vandalism.) New works in, or extensions to, existing buildings must be covered by an 'All Risks' insurance policy (C.2) which, as with Options A and B, must be for the full reinstatement value of the works, including professional fees.

8.19   In cases where the employer may have difficulty in obtaining the joint names insurance for the existing building, which might be the case with tenants and homeowners, the contract now offers an option whereby the parties may 'disapply' paragraph C.1 and replace it with alternative provisions. This must be stated in the contract particulars, which must also describe the document in which the alternative provisions are set out. The employer will need to consider what these provisions might be before tenders are sought, and there are likely to be negotiations before the matter can be finalised. All relevant insurers should, of course, be consulted and (particularly for inexperienced employers) specialist advice may well be required. If the contractor defaults, the employer may take out the insurance and deduct the cost from any sums due or to become due to the contractor, or recover them as a debt (cl 6.12.2).

Action following damage to the works

8.20   The procedure is similar under Insurance Options A, B and C. The contractor must notify the contract administrator and the employer of the details of the damage (cl 6.13.1). The insurers are immediately informed. After any inspection required has been made by the insurers the contractor is then obliged to make good the damage and continue with the works (cl 6.13.4). Under all three options, the contractor authorises the payment of all monies due under the insurance policy to be made directly to the employer (cl 6.13.3).

8.21   Clause 6.13.2 states that 'the occurrence of such loss or damage to executed work or Site Materials shall be disregarded in computing any amounts payable to the Contractor'. Interim certificates that have already been issued and the amounts paid or due under them are, of course, not affected by the occurrence of the damage. In addition, any work that was completed after the most recent interim certificate but was then subsequently damaged should also be included in the next interim payment.

8.22   Under Option A, the contractor must take out insurance for the full reinstatement value of the works, plus a percentage to cover professional fees if this is required in the contract particulars (A.1). In the event of any damage occurring, the insurance money paid to the employer, less the portion of it to cover professional fees, is paid through separate reinstatement work certificates issued by the contract administrator 'at the same dates as those for Interim Certificates under clause 4.9 but without deduction of Retention'. If the amount paid by the insurers is less than it costs the contractor to rebuild the works, the contractor is not entitled to any additional payment (cl 6.13.5). The risk of any underinsurance therefore lies with the contractor.

8.23   Under Options B and C, the rebuilding, restoration or repair work is treated as if it were a variation under clause 3.14 (cl 6.13.6), therefore the contractor is less at risk and the employer would have to bear any shortfall in the monies paid. Under clause 2.29.10, the contractor is entitled to an extension of time for delay caused by loss or damage due to one or more of the 'Specified Perils'. In addition, if the work is treated as a variation under clause 3.14, the contractor may be entitled to an extension of time and loss and/or expense under clauses 2.29.1 and 4.22.1. In all cases, the entitlement appears to extend even to situations where the damage was caused by the contractor's negligence.

8.24   Either party is given the right to terminate the employment of the contractor where there is extensive damage to existing structures, and if 'it is just and equitable' to do so, by means of a notice issued within 28 days of the damage occurring (cl 6.14). The question of termination might arise, for example, where an existing structure to which work is being carried out has been completely destroyed, and it would be unreasonable to expect the contractor to rebuild. If the other party disagrees and feels that the project should continue, it must invoke the dispute resolution procedures. If the contract is terminated, the provisions of clauses 8.12 (except for 8.12.3.5) will apply. It should be noted that this right is in addition to the right under clause 8.11 of either party to terminate the contractor's employment should the works be suspended for a period of two months (or any other period specified in the contract particulars) due to loss or damage caused by any risk covered by the works insurance policy or by any excepted risk (cl 8.11.1.3).

Terrorism cover

8.25   Under clause 6.10, the contractor (where Insurance Option A applies) or the employer (where Insurance Option B or C applies) is required to take out terrorism cover. This can be done either as an extension to the joint names policy or as a separate joint names policy, and must be taken out in the same amount and for the required period of the joint names policy. 'Terrorism Cover' is defined as 'Pool Re Cover or other insurance against loss or damage to work executed and Site Materials (and/or, for the purposes of clause 6.11.1, to an Existing Structure and/or its contents) caused by or resulting from terrorism' (cl 6.8). 'Pool Re Cover' is also a defined term, and stands for 'such insurance against loss or damage to work executed and Site Materials caused by or resulting from terrorism as is from time to time generally available from insurers who are members of the Pool Reinsurance Company Limited scheme or of any similar successor scheme' (cl 6.8). Although there have been difficulties in the past in obtaining such cover, at the time of writing, the insurance market was prepared to cover terrorism risks. If cover is not available, or is likely to differ from the contractual requirements, then, as with insurance generally, the details must be agreed before the contract is signed.

8.26   If the insurers named in the joint names policy decide to withdraw this cover and notify either party, that party must immediately notify the other that terrorism cover has ceased (cl 6.11.1). The employer must then decide whether or not it wishes to continue with the works, and notify the contractor accordingly (cl 6.11.2). If the employer decides to terminate the contractor's employment, the provisions of clauses 8.12.2–8.12.5 apply, excluding clause 8.12.3.5 (cl 6.11.4, see Chapter 9). Otherwise, should any damage be caused by terrorism, then clauses 6.13 and 6.14 apply as discussed above (cl 6.11.5).

Professional indemnity insurance

8.27   The contractor is required to carry professional indemnity insurance (cl 6.15). The level and amount of cover must be inserted in the contract particulars – if no level is inserted it will be 'the aggregate amount for any one period of insurance', and if no amount is stated then no insurance will be required. There is provision for inserting a level of cover for pollution or contamination claims. In addition, if the expiry period is to be 12 years from practical completion then this must be indicated, otherwise the period will be six years. The insurance must be taken out immediately following the execution of the contract, and maintained until the date of practical completion. The contractor must provide evidence of the insurance if required.

The Joint Fire Code

8.28   The Joint Fire Code (cl 6.17–6.20) is designed to reduce the incidence of fire on construction sites. It is an optional provision (cl 6.17) but, as compliance with the code may reduce the cost of some insurance policies, its inclusion should be carefully considered. If it is included, both parties undertake to comply with the code and ensure that those employed by them also comply (cl 6.18).

8.29   If a breach of the Joint Fire Code occurs, the contractor is required to carry out any remedial measures specified in a notice that relate to its obligation to complete the works (cl 6.19.1.1). If the contractor does not comply with the notice within seven days, the employer may employ and pay others to effect such compliance (cl 6.19.2). Where the specified measures require a variation to the works, the contract administrator should issue an instruction (cl 6.19.1.2).

8.30   There are some potential problems with respect to the Joint Fire Code clauses. There is no absolute requirement that the remedial measures required by the insurers must be covered by an instruction of the contract administrator, a situation that could give rise to confusion in practice. Also, there is no reciprocal right for the contractor to employ others, etc. if the employer fails in some way to comply with a requirement of the insurers. This might arise, for example, where persons engaged by the employer fail to comply.

Other insurance

8.31   There remain risks to the employer that are not covered by the SBC16 insurance provisions. For example, if the contractor is caused delay by one of the specified perils, an extension of time would normally be awarded under clause 2.29.10 and the employer will not be able to claim liquidated damages from the contractor for that period. There will therefore be a loss to the employer. Should the employer wish to be insured against this loss of liquidated damages, then special provisions must be made, as there is nothing in SBC16 which deals with such loss. The possible risks should be explained to the employer, but it should be noted that there are often problems with such insurance, as liquidated damages are payable without proof and, traditionally, insurers only pay on proof of actual loss. As a result, only one or two firms are currently willing to offer cover, and the price tends to be high.

8.32   There are other forms of insurance that are not covered by the provisions of SBC16, and which the employer might wish to consider. The employer is the party best placed to assess possible loss. Where there are likely to be business or other economic losses, then these can be covered, albeit at a price. It is also possible to insure against defects occurring in the building by means of project-related insurance. This insurance is still relatively expensive and limited to a ten-year 'decennial' loss. Irrespective of blame, it means that money is available for remedying the defects which will occur most often in the first eight years of the life of a building. Project-related insurance should include for subrogation waiver, and does not reduce the need for professional indemnity cover.

The contract administrator's role in insurance

8.33   The contract administrator has a duty to explain the provisions of the contract to the employer. The choice of the appropriate option for insuring 'the Works' is particularly important and advice must be given to the employer concerning the consequences.

8.34   The employer should take advice from its own insurance experts concerning the suitability and wording of any policies. The contract administrator is primarily a channel of communication, and although a check should be carried out on wording to see that no undesirable exceptions or restrictions exist that might affect the carrying out of the works, the main responsibility should rest with the employer and the employer's broker or insurance advisers.

8.35   Where the insurance requirements of the contract cannot be matched by effective cover, then the employer should seek expert advice. For example, the building might be special and uninsurable, or the employer might not wish to have insurance, etc. Decisions in such situations will also have implications for contractors and sub-contractors, and expert advice must be sought.

9 Termination

9.1   Despite the good intentions of both parties at the outset of a contract, breaches of contract sometimes occur. Some breaches are mere technicalities; some are more serious, although it may be difficult to substantiate damage (e.g. failure to provide a master programme to time); some are so serious that they go to the root of the contract.

9.2   Building contracts usually have express provisions to deal with certain foreseeable situations which might otherwise be breaches. For example, failure to give possession of the site can be dealt with under deferment (cl 2.5), failure to provide information at the agreed dates can be dealt with under extensions of time (cl 2.28), and imposing new restrictions on the contractor's working methods can be dealt with under variations (cl 5.1). These can all therefore be dealt with by the machinery of the contract. If the machinery is not operated as it should be, then the injured party may be able to claim damages for breach of contract. Such claims would have to go to adjudication, arbitration or litigation. For more serious breaches, the contract contains provisions allowing the termination of the employment of the contractor.

Repudiation or termination

9.3   In any contract, where the behaviour of one party makes it difficult or impossible for the other to carry out its contractual obligations, the injured party might allege prevention of performance and sue, either for damages or a *quantum meruit*. This could occur in construction, for example, where the employer refuses to allow the contractor access to part of the site.

9.4   Where it is impossible to expect further performance then the allegation might be that of repudiation. Repudiation occurs when one party makes it clear that it no longer intends to be bound by the provisions of the contract. This intention might be expressly stated, or implied by the party's behaviour.

9.5   JCT contracts include termination clauses, which provide for the effective termination of the employment of the contractor in circumstances which may amount to, or which may fall short of, repudiation. It should be noted that the termination is of the contractor's employment, and is not termination of the contract itself. This means that the parties remain bound by its provisions. If repudiation occurs, it is not necessary to invoke a termination clause because the injured party can accept the repudiation and bring the contract to an end. However, the termination provisions are useful in setting out the exact circumstances, procedures and consequences of the termination of employment. These procedures must be followed with great caution because, if they are not administered strictly in accordance with the terms of the contract, this in itself could amount to a repudiation. This, in turn, might give the other party the right to treat the contract as at an end and claim damages.

9.6   Termination can be initiated by the employer in the event of specified defaults by the contractor such as suspending the works or failing to comply with the CDM Regulations (cl 8.4), or in the event of the insolvency of the contractor (cl 8.5), or in cases of corruption or where certain circumstances relating to the Public Contracts Regulations 2015 apply (cl 8.6). Termination can be initiated by the contractor in the event of specified defaults by the employer, such as failure to pay the amount due on a certificate or where specified events result in the suspension of work beyond a period to be entered in the contract particulars (cl 8.9) or in the event of insolvency of the employer (cl 8.10). Termination might also follow the insolvency of the employer. In the event of neutral causes, which bring about the suspension of the uncompleted works for the period listed in the contract particulars, the right of termination can be exercised by either party (cl 8.11).

Termination by the employer

9.7   The contract provides for termination of the employment of the contractor under stated circumstances (cl 8.4). SBC16 expressly states that the right to terminate the contractor's employment is 'without prejudice to any other rights and remedies' (cl 8.3.1). This

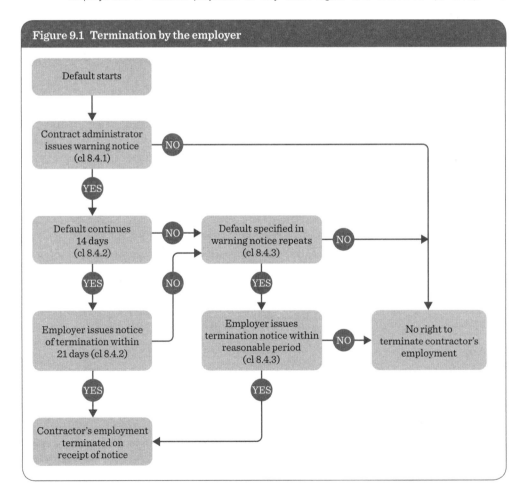

Figure 9.1 Termination by the employer

termination can be initiated by the employer in the event of specified defaults by the contractor occurring prior to practical completion (cl 8.4), the insolvency of the contractor (cl 8.5) or corruption (cl 8.6). Where the employer is a local or public authority, circumstances as set out in regulation 73(1)(b) of the Public Contracts Regulations 2015 (conviction of various offences) will also give rise to the right to terminate (cl 8.6).

9.8   The procedures as set out in the contract must be followed exactly, especially those concerning the issue of notices under clause 8.4 (see Figure 9.1). If default occurs, the contract administrator should issue a warning notice, which should specify the particular default (although not necessarily a detailed list of circumstances) (cl 8.4.1). If the default continues for 14 days from receipt of the notice, then the employer may terminate the employment of the contractor by the issue of a further notice within 21 days from the expiry of the 14 days (cl 8.4.2). If the contractor ends the default or if the employer gives no further notice, and the contractor then repeats the default, the employer may terminate 'within a reasonable time after such repetition' (cl 8.4.3). The employer must still give a notice of termination, but no further warning is required from the contract administrator. There appears to be no time limit on the repetition of the default. If, however, a considerable period has elapsed, it may be prudent for the employer to issue a further warning notice before issuing the notice of termination.

9.9   It should be noted that, to be valid, all notices must be given strictly in accordance with clause 1.7.4, i.e. 'by hand or sent by Recorded Signed for or Special Delivery post' (cl 8.2.3). It should be noted that this is not the same as the older wording 'actual delivery' so it is unlikely that fax or email would be acceptable, as it was in the case of *Construction Partnership* v *Leek Developments*. As time limits are of vital importance here, it is usually wise to have receipt of delivery confirmed.

> **Construction Partnership UK Ltd v Leek Developments Ltd [2006] CILL 2357 TCC**
>
> On an IFC98 contract, a notice of determination was delivered by fax, but not by hand or by special delivery or recorded delivery. (A letter had been sent by normal post but it was unclear whether or not it had been received.) Clause 7.1 required actual delivery of notices of default and determination, and the contractor disputed whether the faxed notice was valid. The court had therefore to decide what 'actual delivery' meant. It decided that it meant what it says: 'Delivery simply means transmission by an appropriate means so that it is received'. In this case, it was agreed that the fax had been received, therefore the notice complied with the clause. The CILL editors state that 'on a practical level, this judgment is quite important' because it had previously been assumed that 'actual delivery' meant physical delivery by hand. In their view, email could be considered an appropriate method of delivery, although that was not decided in the case.

9.10   The grounds for termination by the employer must be clearly established and expressed. The contract clearly states that termination must not be exercised unreasonably or vexatiously (cl 8.2.1). Before issuing any notice the contract administrator should check, for example, that all extensions of time have been dealt with in accordance with the contract. Under clause 8.4.1.1, suspension of the work must be whole or substantial, and 'without reasonable cause'. However, the contractor might find 'reasonable cause' in any of the matters referred to in clause 8.9.1. An exercise of the contractor's right to suspend work (HGCRA 1996) would not be cause for termination, provided that it had been exercised in accordance with the terms of the contract.

9.11   The 'specified defaults' which may give rise to termination are that the contractor:

- wholly or substantially suspends the design or construction of the works (cl 8.4.1.1);
- 'fails to proceed regularly and diligently' with the performance of its obligations (cl 8.4.1.2);
- refuses or neglects to comply with a written notice or instruction requiring the contractor to remove defective work (cl 8.4.1.3);
- fails to comply with clause 3.7 (sub-contracting) or clause 7.1 (assignment) (cl 8.4.1.4);
- fails to comply with clause 3.23 (CDM Regulations) (cl 8.4.1.5).

9.12   Generally speaking, only a serious default would justify termination, although any failure to comply with the CDM Regulations' provisions which would put the employer at risk of action by the authorities would be sufficient.

9.13   The default that the contractor 'fails to proceed regularly and diligently' is notoriously difficult to establish. Careful records kept by the employer's representative (e.g. a clerk of works) or the contract administrator will be of utmost importance should the contractor dispute this matter. 'Default' here means more than simply falling behind any submitted programme (relating to design and/or construction work), even to such an extent that it is quite clear the project will finish considerably behind time. However, something less than a complete cessation of construction work would be sufficient grounds.

9.14   In the case of *London Borough of Hounslow* v *Twickenham Garden Developments*, for example, the architect's notice was strongly attacked by the defendants. In a more recent case, however, the contract administrator was found negligent because it failed to issue a notice (*West Faulkner Associates* v *London Borough of Newham*). It should be remembered that without the first 'warning notice' issued by the contract administrator the employer cannot issue the termination notice.

London Borough of Hounslow v *Twickenham Garden Developments* (1970) 7 BLR 81

The London Borough of Hounslow entered into a contract with Twickenham Garden Developments to carry out sub-structure works at Heston and Isleworth in Middlesex. The contract was on JCT63. Work on the contract stopped for approximately eight months due to a strike. After work resumed, the architect issued a notice of default stating that the contractor had failed to proceed regularly and diligently and that, unless there was an appreciable improvement, the contract would be determined. The employer then proceeded to determine the contractor's employment. The contractor disputed the validity of the notices and the determination, and refused to stop work and leave the site. Hounslow applied to the court for an injunction to remove the contractor. The judge emphasised that an injunction was a serious remedy and that, before he could grant one, there had to be clear and indisputable evidence of the merits of Hounslow's case. The evidence put before him, which showed a significant drop in the amounts of monthly certificates and numbers of workers on site, failed to provide this.

West Faulkner Associates v *London Borough of Newham* (1992) 61 BLR 81

West Faulkner Associates were architects engaged by the London Borough of Newham for the refurbishment of a housing estate consisting of several blocks of flats. The residents of the estate were evacuated from their flats in stages to make way for the contractor, Moss, which, it had been agreed, would carry out the work according to a programme of phased possession and completion, with each block taking nine weeks. Moss fell behind the programme almost immediately. However, Moss had a large workforce on the site and continually promised to revise its programme and working methods to address the problems of lateness, poor quality work and unsafe working practices that were drawn to its attention on numerous occasions by the architect. In reality, Moss remained completely disorganised, and there was no apparent improvement. The architect took the advice of quantity surveyors that the grounds of failing to proceed regularly and diligently would be difficult to prove, and decided not to issue a notice. As a consequence, Newham was unable to issue a notice of determination, had to negotiate a settlement with the contractor and dismissed the architect, which then brought a claim for its fees.

The judge decided that the architect was in breach of contract in failing to give proper consideration to the use of the determination provisions. In his judgment, he stated that 'regularly and diligently' should be construed together and in essence they mean simply that the contractors must go about their work in such a way as to achieve their contractual obligations. 'This requires them to plan their work, to lead and manage their workforce, to provide sufficient and proper materials and to employ competent tradesmen, so that the works are carried out to an acceptable standard and that all time, sequence and other provisions are fulfilled' (Judge Newey at page 139).

Insolvency of the contractor

9.15   Insolvency is the inability to pay debts as they become due for payment. Insolvent individuals may be declared bankrupt. Insolvent companies may be dealt with in a number of ways, depending upon the circumstances: for example, by voluntary liquidation (in which the company resolves to wind itself up); compulsory liquidation (under which the company is wound up by a court order); administrative receivership (a procedure to assist the rescue of a company under appointed receivers); an administration order (a court order given in response to a petition, again with the aim of rescue rather than liquidation, and managed by an appointed receiver); or voluntary arrangement (in which the company agrees terms with creditors over payment of debts). SBC16 sets out a full definition of the term 'insolvent' for the purposes of the contract at clause 8.1. Procedures for dealing with insolvency are mainly subject to the Insolvency Act 1986 and the Insolvency Rules 2016 (in place from April 2017). Under these provisions, the person authorised to oversee statutory insolvency procedures is termed an 'insolvency practitioner'.

9.16   Under SBC16, the contractor must notify the employer in writing in the event of liquidation or insolvency (cl 8.5.2). The employer is given an option to terminate (cl 8.5.1) or to consider a more constructive approach. This is to allow the appointed insolvency practitioner time to come up with a rescue package if possible. It is usually in the employer's interest to have the works completed with as little additional delay and cost as possible, and a breathing space might allow all possibilities to be explored. During this period, the contract states that 'clauses 8.7.3 to 8.7.5 and (if relevant) clause 8.8 shall apply as if such notice had been given' (cl 8.5.3.1). This means that even if no notice of termination is given, the employer is under no obligation to make further payment except as provided under those clauses (see paragraph 9.23). The contractor's obligation to 'carry out and complete the Works and the design of the Contractor's Designed Portion' is suspended (cl 8.5.3.2), and the employer may take reasonable measures to see that the site and works are protected (cl 8.5.3.3). The employer can then either make an agreement to arrange for the work to continue or terminate the employment of the contractor.

9.17   There are several options for completing the project. The first allows for arrangements to be made for the contractor to continue and complete the works. Unless the insolvency practitioner has been able to arrange resource backing, this may not be a realistic option. If practical completion is near, however, and there is money due to the contractor, it can be advantageous to allow completion under the control of the insolvency practitioner.

9.18   Alternatively, another contractor may be novated to complete the works. On a 'true novation', the substitute contractor takes over all the original obligations and benefits (including completion to time and within the contract sum). More likely is the third option, 'conditional novation', whereby the contract completion date, etc. would be subject to renegotiation, and the substitute contractor would probably want to disclaim liability for that part of the work undertaken by the original contractor.

9.19   Deciding on which of the options would best serve the interests of all the parties is a matter to be resolved between the employer, the contract administrator and the insolvency practitioner. A pragmatic approach might be to continue initially with the original contractor under an interim arrangement until such time as novation can be arranged or a completion contract negotiated.

Consequences of termination

9.20   If the employer exercises its right to terminate under clause 8.4, 8.5 or 8.6, then the only way to achieve completion will be through the appointment of a new contractor of the employer's choice. The contract gives the employer the right to employ others under clause 8.7.1, to both carry out the works and complete the contractor's designed portion. A completion contract might result from negotiation or competitive tender. The employer will have the right to use any temporary buildings, plant, etc. on site, including those which are not owned by the original contractor, subject to the consent of the owner (cl 8.7.1).

9.21   The employer may also require the contractor to:

- remove from the site any temporary buildings, plant, etc. which are owned by the contractor (cl 8.7.2.1);
- provide the employer with copies of all contractor's design documents (cl 8.7.2.2);
- require the original contractor to assign the benefit of any sub-contracts to the employer (to the extent that the benefit is assignable) (cl 8.7.2.3).

9.22   If the employer decides to employ others under clause 8.7.1, such employment must be handled with care, as completion of a building started by another contractor is always difficult. A completion contract might result from negotiation or competitive tender, but the tender route may be advisable if there is much to complete as the employer may have to demonstrate subsequently that the costs incurred were reasonable. A record should be made of the exact state of completeness at the time of termination, including any defective work.

9.23   Following termination, clause 8.7.3 states that 'no further sum shall become due to the Contractor … other than any amount that may become due to him under clause 8.7.5 or 8.8.2' (i.e. the payment provisions following termination). It also states that the employer will not need to make any payments that have already become due to the extent that a pay less notice has been given (cl 8.7.3.1) or where the contractor has become insolvent (cl 8.7.3.2). This reflects section 111(10) of the HGCRA 1996 as amended, and the judgment in *Melville Dundas* v *George Wimpey*. It should be noted, however, that the employer may still be obliged to pay amounts awarded by an adjudicator (*Ferson* v *Levolux*).

Melville Dundas Ltd v George Wimpey UK Ltd [2007] 1 WLR 1136 (HL)

On a contract let on WCD98, the contractor had gone into receivership, entitling the employer to determine the contractor's employment. The contractor had applied for an interim payment on 2 May 2003, the final date for payment was 16 May (14 days after application), and the determination was effective on 30 May 2003. The contractor claimed the payment on the basis that no withholding notice had been issued. By a majority of three to two, the House of Lords decided that the employer was not obliged to make any further payment. It was accepted that, under WCD98, interim payments were not contractually payable after determination and the House of Lords held that this was not inconsistent with the payment provisions of the HGCRA 1996. Although the Act requires that the contractor should be entitled to payment in the absence of a notice, this did not mean that that entitlement had to be maintained after the contractor had become insolvent, i.e. it was not inconsistent to construe that the effect of the determination was that the payment was no longer due. The Act was concerned with the balance of interests between payer and payee, and to construe it otherwise would give a benefit to the contractor's creditors against the interests of the employer, something which the Act did not intend.

Ferson Contractors Ltd v Levolux A T Ltd [2003] BLR 118

Ferson was the contractor and Levolux the sub-contractor on a GC/Works sub-contract. A dispute arose regarding Levolux's second application for payment; £56,413 was claimed but only £4,753 was paid. A withholding notice was issued which specified the amount but not the reason for withholding it. Levolux brought a claim to adjudication, and the adjudicator decided that the notice did not comply with section 111 of the HGCRA 1996, and that Ferson should pay the whole amount. Ferson refused to pay and Levolux sought enforcement of the decision. Prior to the adjudication, Levolux had suspended work and Ferson, maintaining that the suspension was unlawful, had determined the contract. Ferson now maintained that, due to clause 29, which stated that 'all sums of money that may be due or accruing due from the contractor's side to the sub-contractors shall cease to be due or accrue due', it did not have to pay this amount. The CA upheld the decision of the judge of first instance that the amount should be paid: 'The contract must be construed so as to give effect to the intent of Parliament'.

9.24   Following termination, a notional final account must be set out, stating what is owed or owing, either in a statement prepared by the employer or in a contract administrator's certificate (cl 8.7.4). This account must be prepared within three months of 'the completion of the Works and the making good of defects in them', which allows the employer a period to assess its losses due to the termination. The net amount shown on the account should be paid by the contractor to the employer (the more likely outcome), or by the employer to the contractor, as appropriate (cl 8.7.5).

9.25   One of the consequences of termination is that it often takes time for the contractor to effect an orderly withdrawal from site, and for the employer to establish the amounts outstanding before final payment. Should the employer decide not to continue with the construction of the works after termination, the employer is required to notify the contractor in writing within six months of the termination of the contractor's employment (cl 8.8.1). Within a reasonable time of the notification (or within six months of termination, if no work is carried out and no notice issued) the employer must send the contractor a statement of the value of the works and losses suffered as required under clause 8.8.

Termination by the contractor

9.26   The contractor has a reciprocal right to terminate its own employment under clause 8.9 in the event of specified defaults of the employer (8.9.1) or specified suspension events (cl 8.9.2) or insolvency of the employer (cl 8.10). The specified suspension events must have resulted in the suspension of the whole of the uncompleted works for the continuous period stated in the contract particulars (if no period is entered, the default period is two months). In the case of specified defaults or suspension events a notice is required, which must specify the default or event. If the default or event continues for 14 days from receipt of the notice, the contractor may terminate the employment by a further notice up to 21 days from the expiry of the 14 days (cl 8.9.3). Alternatively, if the employer ends the default or the suspension event ceases, and the contractor gives no further notice, should the employer repeat the default the contractor may terminate 'within a reasonable time after such repetition' (cl 8.9.4). As for the employer, these notices must be given by the means set out in clause 1.7.4, i.e. 'by hand or sent by Recorded Signed for or Special Delivery post' (cl 8.2.3).

9.27   The grounds of clause 8.9 differ from those that give the employer the right to terminate. They comprise failure to pay the amount due to the contractor in accordance with clause 4.11 (cl 8.9.1.1), obstruction of the issue of a certificate (cl 8.9.1.2), failure to comply with the contractual provisions relating to assignment under clause 7.1 (cl 8.9.1.3) or with CDM obligations under clause 3.23 (cl 8.9.1.4). There are also matters which relate directly to the duties of the contract administrator, where, for example, the carrying out of the whole or substantially the whole of the works is suspended for a period stated in the contract particulars due to contract administrator's instructions under clauses 2.15, 3.14 or 3.15 (cl 8.9.2.1), or due to 'any impediment, prevention or default' of the employer (cl 8.9.2.2). The period of suspension is two months unless some other period is stated. In long and complex projects two months may not be sufficient should unexpected technical problems arise, therefore it may be advisable to consider inserting a longer period.

9.28   The contractor should exercise particular care if considering termination due to what it considers to be an employer's failure to pay. The contract specifically requires failure to pay 'the amount due' (cl 8.9.1.1), and not simply the amount for which the contractor might have applied. If the contractor has made an error in its calculations, the employer might be entitled to pay a lesser amount. In addition, the contract gives the employer rights to make various adjustments to and deductions from amounts due. The correct exercise of these rights would not amount to a failure to pay an amount due. If the contractor attempts to terminate the contract without justification, this will amount to repudiation, with serious consequences for the contractor. A more prudent course would be to raise the disputed payment in adjudication, while continuing to proceed with the works.

9.29   Termination by the contractor is optional in the case of the employer's bankruptcy or insolvency (cl 8.10.1). The contractor must issue a notice and termination would take effect from the receipt of the notice.

Consequences of termination by the contractor

9.30   The contractor must remove from the site all temporary buildings, tools, etc., and provide the employer with all contractor design documents and related information (cl 8.12.2). The contractor then prepares an account setting out the total value of the work at the date of termination, plus other costs relating to the termination as set out in clause 8.12.3. These may include such items as the cost of removal and any direct loss and/or damage consequent upon termination (cl 8.12.3.3 and 8.12.3.5). The contractor is, in effect, indemnified against any damages that may be caused as a result of the termination. This would not necessarily be the case if the contractor were not to comply with the contractual provisions; in that case the contractor's actions might constitute repudiation.

Termination by either the employer or the contractor

9.31   Either party is given the right to terminate if the carrying out of the works is wholly or substantially suspended for the continuous period stated in the contract particulars due to one or more of the events listed in clause 8.11.1 (if no period is entered, the default period is two months). These events include force majeure, loss or damage to the works caused by any risk covered by the works insurance policy or by an excepted risk, civil commotion and the exercise by the government or a local or public authority of a statutory power not occasioned by the default of the contractor. The right of the contractor to terminate in the event of damage to the works is limited by the proviso that the event must not have been caused by the contractor's negligence (cl 8.11.2). In addition, either party may terminate if work is suspended because of an employer's instruction under clause 2.15 (discrepancies), clause 3.14 (variations) or clause 3.15 (postponement) which has been issued as a result of negligence or default of a statutory undertaker (cl 8.11.1.2).

9.32   Notice may be given by either party and the employment of the contractor will be terminated seven days after receipt of the notice, unless the suspension is terminated within that period (cl 8.11.1). If work is not resumed after this period, the party may then, by further notice, terminate the contract (cl 8.11.1). Where the employer is a local or public authority, the employer may issue a notice if circumstances in regulation 73(1)(a) or (c) of the Public Contracts Regulations 2015 (various breaches of the Regulations) apply (cl 8.11.3). This appears to effect an immediate termination.

9.33   Detailed provisions are set out regarding the consequences of the termination. Clause 8.12.1 states that 'no further sums shall become due to the Contractor otherwise than in accordance with this clause', which in effect means that other provisions of the contract requiring further payment will cease to operate. The contractor must remove all temporary buildings, tools, etc. from the site (cl 8.12.2). An account is then prepared in the same format as for termination by the contractor (cl 8.12.3, see paragraph 9.30 above), except that, in this case, it may include amounts relating to direct loss and/or damage to the contractor resulting from a specified peril caused by the employer's negligence.

Termination of a named specialist's employment

9.34   Where a named specialist has been appointed under Supplemental Provision 9, the contractor may only terminate its employment if it has first notified and consulted with the contract administrator (Schedule 8:9.5.1). The notification must be given promptly, as soon as the contractor has the right to terminate the appointment or to treat it as repudiated, Following this notification, except in cases where the sub-contractor has become insolvent, it must not give notice of termination to the named specialist until at least 14 days have elapsed (Schedule 8:9.5.2), and must send copies of the notice to the contract administrator (Schedule 8:9.5.3).

9.35   If such termination takes place, the contract administrator must give instructions that either select another named specialist, direct the contractor to carry out the work or omit the named specialist work from the contract (Schedule 8:9.6 and 8:9.3). If no instruction is given within seven days of receiving the copy of the contractor's termination notice, the contract administrator will be deemed to have instructed the contractor to complete the work. These instructions are subject to the rights of objection, and would have the same contractual implications, as discussed at paragraph 5.69.

10 Dispute resolution

10.1   SBC16 refers to five methods of dispute resolution: negotiation; mediation; adjudication; arbitration; and legal proceedings. One of these, adjudication, is a statutory right, and if one party wishes to use this method, the other must concur. Negotiation is an optional provision (Schedule 8, Supplemental Provision 6). Negotiation and mediation are voluntary processes which depend on the co-operation of the parties, and either may lead to a binding result. If none of the options of negotiation, mediation or adjudication is used, or if either party is dissatisfied with the decision of an adjudicator, then the dispute will have to be resolved by arbitration or litigation.

10.2   SBC16 requires the parties to decide in advance whether arbitration or litigation will be used. If arbitration is to be the final method of dispute resolution, then the contract particulars must indicate that 'Article 8 and clauses 9.3 to 9.8 apply'. If this is the case, then all disputes will be referred to arbitration, except for those relating to the Construction Industry Scheme, VAT or the enforcement of an adjudicator's decision.

10.3   There are therefore stages, either before or during the contract, where the parties have the opportunity to agree a preferred course of action. It is important for the contract administrator to understand and to be able to advise on these methods. The contract administrator may also be asked to give evidence, so a basic understanding of the procedures involved is essential.

Negotiation

10.4   If Supplemental Provision 6 is incorporated, the parties are each obliged to notify the other promptly of any matter that may give rise to a dispute. The senior executives nominated in the contract (or persons of equal standing) must then meet and, in good faith, try to resolve the matter. There is no sanction for non-compliance, but refusal or failure to comply might be taken into account in any subsequent legal proceedings. Even if the provision is not incorporated, before any of the more formal procedures are initiated, there may be a period of negotiation where the parties attempt to resolve their differences themselves. This might be the best solution to the problem, but the contract administrator should tread carefully if considering becoming involved in such negotiations. The contract administrator may be of great assistance in advising the employer and providing information, but has no authority to negotiate amendments to the terms of the contract or make ad hoc agreements on behalf of the employer. Even if the employer gives the contract administrator an extended authority to negotiate a settlement, where the dispute involves complex legal points, a lawyer would be the best choice to handle the negotiations.

Alternative dispute resolution

10.5   If negotiations fail to achieve an agreement, the parties may submit the dispute to 'alternative dispute resolution' (ADR), a term used to cover methods such as conciliation, mediation and the mini-trial. SBC16 clause 9.1 requires each party to give serious consideration to a request by the other to use mediation. A footnote to clause 9.1 refers to the JCT Guide (SBC/G). The Guide does not itself set out or advocate any particular procedure to be used in mediation; instead, it states that such choices are frequently better made by the parties when the dispute has actually arisen. The parties could, of course, supplement SBC16 by selecting a procedure or mediator appointing body and setting this out in the terms of their appointment. As mediation is a consensual process, any individual reference to mediation would have to be supported by both parties.

10.6   Usually a mediator is appointed jointly by the parties and will meet with the parties together and separately in an attempt to resolve the differences. The outcome is often in the form of a recommendation which, if acceptable, can be signed as a legally binding agreement. This would then be enforceable in the same way as any other contract. However, if the recommendation is not acceptable to one of the parties and is not signed as a binding agreement, it cannot be imposed by law, and so the time spent on the mediation may appear to have been wasted.

10.7   Nevertheless, there can be many advantages to mediation. Unlike adjudication, arbitration or litigation, it is a non-adversarial process which tends to forge good relationships between the parties. Imposed solutions may leave at least one of the parties dissatisfied and may make it very difficult for the parties to work together in the future. If the parties are keen to promote a long-term business relationship they should give mediation serious consideration. Even if mediation does not result in a complete solution, it has been found in practice that it can help to clear the air on some of the issues involved and to establish common ground. This, in turn, might then pave the way for shorter and possibly less acrimonious arbitration or litigation.

Adjudication

10.8   The Housing Grants, Construction and Regeneration Act (HGCRA) 1996 Part II requires that parties to the construction contracts falling within the definition set out in the Act have the right to refer any dispute to a process of adjudication which complies with requirements stipulated in the Act. Article 7 of SBC16 restates this right, and refers to clause 9.2, which states that where a party decides to exercise this right 'the Scheme shall apply'. This refers to the Scheme for Construction Contracts, a piece of secondary legislation which sets out a procedure for the appointment of the adjudicator and the conduct of the adjudication. The Scheme takes effect as implied terms in a contract, if and to the extent that the parties have failed to agree on a procedure that complies with the Act.

10.9   By stating 'the Scheme shall apply', SBC16 is effectively annexing the provisions of the Scheme to the form, which therefore become a binding part of the agreement between the parties. Clause 9.2, however, makes its application subject to certain conditions which relate to the appointment of the adjudicator.

10.10   Under SBC16 the adjudicator may either be named in the contract particulars or nominated by the nominating body identified in the contract particulars. A named adjudicator will normally enter into the JCT Adjudication Agreement (Named Adjudicator) (Adj/N) with the parties at the time the main contract is entered into.

10.11   The party wishing to refer a dispute to adjudication must first give notice under paragraph 1(1) of the Scheme (see Figure 10.1). The notice may be issued at any time and should

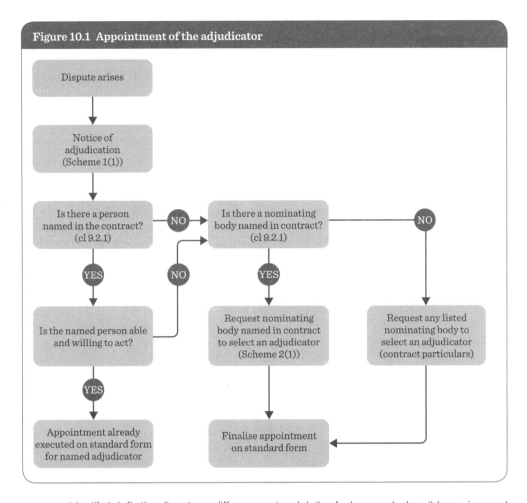

Figure 10.1 Appointment of the adjudicator

identify briefly the dispute or difference, give details of where and when it has arisen, set out the nature of the redress sought, and include the names and addresses of the parties, including any specified for the giving of notices (paragraph 1(3)). If no adjudicator is named, the parties may either agree an adjudicator or either party may apply to the 'nominator' identified in the contract particulars (paragraph 2(1)). If no nominator has been selected, then the contract states that the referring party may apply to any of the nominators listed in the contract particulars. The adjudicator will then send terms of appointment to the parties. In addition to the form for a named adjudicator, the JCT also publishes the Adjudication Agreement (Adj) for use in this situation.

10.12   The Scheme does not stipulate any qualifications in order to be an adjudicator, but does state that the adjudicator 'shall be a natural person acting in his personal capacity' and should not be an employee of either of the parties (paragraph 4). In addition, SBC16 requires that, where the dispute relates to clause 3.18.4 (repeat testing), the person appointed shall 'where practicable' be 'an individual with appropriate expertise and experience in the specialist area or discipline relevant to the instruction or issue in dispute' (cl 9.2.2.1). Where the person appointed does not have the appropriate expertise, that person must appoint an independent expert to advise and report.

10.13   The adjudicator is required to act impartially, must avoid incurring unnecessary expense (paragraph 12), and is not liable for anything done or omitted when acting properly as an adjudicator (paragraph 26).

10.14   The referring party must refer the dispute to the selected adjudicator within seven days of the date of the notice (paragraph 7(1)). The referral will normally include particulars of the dispute, and must include a copy of, or relevant extracts from, the contract and any material it wishes the adjudicator to consider (paragraph 7(2)). A copy of the referral must be sent to the other party and the adjudicator must inform all parties of the date it was received (paragraph 7(3)).

10.15   The adjudicator will then set out the procedure to be followed. A preliminary meeting may be held to discuss this, otherwise the adjudicator may send the procedure and timetable to both parties. The party which did not initiate the adjudication (the responding party) will be required to respond by a stipulated deadline. The adjudicator is likely to hold a short hearing of a few days at which the parties can put forward further arguments and evidence. There may also be a site visit. Occasionally it may be possible to carry out the whole process by correspondence (often termed 'documents only').

10.16   The adjudicator is given considerable powers under the Scheme (e.g. paragraphs 13 and 20), including the right to take the initiative in obtaining the facts and the law, the right to issue directions, the right to revise decisions and certificates of the contract administrator, the right to carry out tests (subject to obtaining necessary consents), and the right to obtain from others necessary information and advice. The adjudicator must give advance notice if intending to take legal or technical advice.

10.17   The HGCRA 1996 requires that the decision is reached within 28 days of referral, but it does not state how this date is to be established (section 108(2)(c)). Under the Scheme, the 28 days starts to run from the date of receipt of the referral notice (paragraph 19(1)). The period can be extended by up to 14 days by the referring party, and further by agreement between the parties. The decision must be delivered forthwith to the parties, and the adjudicator may not retain it pending payment of the fee. The provisions state that the adjudicator must give reasons for the decision if requested to do so by the parties (paragraph 22).

10.18   The parties must meet their own costs of the adjudication, unless they have agreed that the adjudicator shall have the power to award costs. Under the Act, any agreement is ineffective unless it complies with section 108A, including that it is made in writing after a notice of adjudication is issued (SBC16 therefore does not contain such an agreement). The adjudicator, however, is entitled to charge fees and expenses (subject to any agreement to the contrary), although expenses are limited to those 'reasonably incurred' (paragraph 25). The adjudicator may apportion those fees between the parties, and the parties are jointly and severally liable to the adjudicator for any sum which remains outstanding following the adjudicator's determination. This means that in the event of default by one party, the other party becomes liable to the adjudicator for the outstanding amount.

10.19   The adjudicator's decision will be final and binding on the parties 'until the dispute is finally determined by legal proceedings, by arbitration, or by agreement between the parties'. The effect of this is that if either party is dissatisfied with the decision, it may raise the dispute again in arbitration or litigation as indicated in the contract particulars, or it may negotiate a fresh agreement with the other party. In all cases, however, the parties remain bound by the decision and must comply with it until the final outcome is determined.

10.20   If either party refuses to comply with the decision, the other may seek to enforce it through the courts. Generally, actions regarding adjudicators' decisions have been dealt with promptly by the courts and the recalcitrant party has been required to comply. Paragraph 22A of the Scheme allows the adjudicator to correct clerical or typographical errors in the decision, within five days of it being issued, either on the adjudicator's own initiative or because the parties have requested it, but this would not extend to reconsidering the substance of the dispute.

Arbitration

10.21   Arbitration refers to proceedings in which the arbitrator has power derived from a written agreement between the parties to a contract, and which is subject to the provisions of the Arbitration Act 1996. Arbitration awards are enforceable at law. An arbitrator's award can be subject to appeal on limited grounds.

10.22   If arbitration is preferred to litigation as the method for final determination of disputes, then this is confirmed by indicating in the contract particulars that Article 8 will apply (note that if no entry is made, the default process will be litigation). The arbitration provisions are set out in clauses 9.3–9.8, which refer to the Construction Industry Model Arbitration Rules (CIMAR; 'the Rules'). The Arbitration Act 1996 confers wide powers on the arbitrator unless the parties have agreed otherwise, but leaves detailed procedural matters to be agreed between the parties or, if not so agreed, to be decided by the arbitrator. To avoid problems arising, it is advisable to agree as much as possible of the procedural matters in advance, and SBC16 does this by incorporating the Rules, which are very clearly written and self-explanatory. The specific edition referred to is the 2016 edition published by the JCT, which incorporates supplementary and advisory procedures, some of which are mandatory (Part A) and some of which apply only if agreed after the arbitration is begun (Part B). The paragraphs below refer to the JCT edition of the Rules.

10.23   The party wishing to refer the dispute to arbitration must give notice as required by SBC16 clause 9.4 and Rule 2.1, identifying briefly the dispute and requiring the party to agree to the appointment of an arbitrator. If the parties fail to agree within 14 days, either party may apply to the 'appointor', selected in advance from a list of organisations set out in the contract particulars. If no appointor is selected, then the contract states that the appointor will be the president or a vice-president of the RIBA. Under Rule 2.5 the arbitrator's appointment takes effect when he or she agrees to act, and is not subject to first reaching agreement with the parties on matters such as fees.

10.24   The arbitrator has the right and the duty to decide all procedural matters, subject to the parties' right to agree any matter (Rule 5.1). Within 14 days of appointment the parties must each send the arbitrator and each other a note indicating the nature of the dispute and amounts in issue, the estimated length for the hearing, if necessary, and the procedures to be followed (Rule 6.2). The arbitrator must hold a preliminary meeting within 21 days of appointment to discuss these matters (Rule 6.3). The first decision to make is whether Rule 7 (short hearing), Rule 8 (documents only) or Rule 9 (full procedure) is to apply. The decision will depend on the scale and type of dispute.

10.25   Under all three Rules referred to above, the parties exchange statements of claim and of defence, together with copies of documents and witness statements on which they intend to rely. Under Rule 8, the arbitrator makes the award based on the documentary evidence only. Under Rule 9, the arbitrator will hold a hearing at which the parties or their representatives can put forward further arguments and evidence. There may also be a site visit. The JCT amendments set out time limits for these procedures.

10.26   Under Rule 7 a hearing is to be held within 21 days of the date when Rule 7 becomes applicable, and the parties must exchange documents not later than seven days prior to the hearing. The hearing should last no longer than one day. The arbitrator publishes the award within one month of the hearing. The parties bear their own costs.

10.27   The arbitrator is given a wide range of powers under the Rules, including the power to obtain advice (Rule 4.2), the powers set out in section 38 of the Arbitration Act 1996 (Rule 4.3), the power to order the preservation of work, goods and materials even though they are a part of work that is continuing (Rule 4.4), the power to request the parties to carry out tests (Rule 4.5), the power to award security for costs (Rule 4.6), and the power to award costs (Rule 13.1). Under clause 9.5 of SBC16 the arbitrator is also given wide powers to review and revise any certificate, opinion, decision, requirement or notice, and to disregard them if need be, where seeking to determine all matters in dispute.

10.28   Costs are normally awarded on a judicial basis, i.e. the loser will pay the winner's costs (Rule 13.1). The arbitrator will be entitled to charge fees and expenses and will apportion those fees between the parties on the same basis. The parties are jointly and severally liable to the arbitrator for fees and expenses incurred.

Arbitration and adjudication

10.29   Under Article 8 any dispute that has been referred to an adjudicator may be referred to arbitration if this is required by either party. The conclusive effect of a final certificate is suspended where adjudication, arbitration or litigation is commenced within 28 days of the date of its issue (cl 1.9.2.1), until the proceedings are concluded. Furthermore, where a party wishes to refer to arbitration a dispute that was the subject of an adjudicator's decision, the conclusive effect will remain suspended, provided the dispute is referred within 28 days of the adjudicator's decision (cl 1.9.2.2).

Arbitration or litigation

10.30   SBC16 contains alternative provisions for arbitration and litigation in Articles 8 and 9, and a choice has to be made before tender documents are sent out. Both processes give rise to binding and enforceable decisions. Both tend to be lengthy and expensive, although there are provisions for short forms of arbitration.

10.31   Litigation cases involving claims for amounts greater than £25,000 are normally heard in the High Court, and construction cases are usually heard in the Technology and Construction Court, a specialist department of the High Court which deals with technical or scientific cases. Procedures in court follow the Civil Procedure Rules, with the timetable and other detailed arrangements being determined by the court. A judge will hear the case and although, in the past, parties were required to be represented by barristers, now they may represent themselves, or elect to be represented by an 'advisor'.

10.32   Disputes in building contracts have traditionally been settled by arbitration. Arbitrators are usually senior and experienced members of one of the construction professions, and for many years it was felt that they had a greater understanding of construction projects and the disputes that arise than might be found in the courts. These days, however, the judges of the Technology and Construction Court have extensive experience of technical construction disputes. The high standards now evident in these courts are likely to be matched in practice by only a few arbitrators.

10.33   The court has powers to order that actions regarding related matters are joined (for example, where disputes between an employer and contractor, and contractor and nominated sub-contractor, concern the same issues). This is much more difficult to achieve in arbitration. Even if all parties have agreed to the use of CIMAR, the appointing bodies must have been alerted and have agreed to appoint the same arbitrator (Rules 2.6 and 2.7). If the same arbitrator is appointed, he or she may order concurrent hearings (Rule 3.7), but may only order consolidated proceedings with the consent of all the parties (Rule 3.9), which is often difficult to obtain. The court's powers may therefore offer an advantage in multi-party disputes, by avoiding duplication of hearings and potentially conflicting outcomes.

10.34   There remain, however, two key advantages to using arbitration. The first is that in arbitration the proceedings can be kept private, which is usually of paramount importance to construction professionals and companies, and is often a deciding factor in selecting arbitration. In court, the proceedings are open to the public and the press, and the judgment is published and widely available.

10.35   The second advantage to the parties is that the arbitration process is consensual. The parties are free to agree on timing, place, representation and the individual arbitrator. This autonomy carries with it the benefits of increased convenience and possible savings in time and expense. The parties avoid having to wait their turn at the High Court and may choose a time and place for the hearing which is convenient to all. In arbitration, however, the parties have to pay the arbitrator and meet the cost of renting the premises in which the hearing is held.

10.36   It should perhaps be noted that even where parties have selected arbitration under Article 8, it is still open for them to elect litigation once a dispute develops. If, however, one party commences court proceedings, the other may ask the court to stay the proceedings on the grounds that an arbitration agreement already exists. (This would not apply to litigation to enforce an adjudicator's decision, as Article 8 excludes all disputes regarding the enforcement of a decision of an adjudicator from the jurisdiction of the arbitrator.) If, on the other hand, the parties had originally selected litigation, this would not prevent them from subsequently agreeing to take a dispute to arbitration, but in such cases they would also have to agree which procedural rules are to apply.

Appendix A: Specimen activity schedule

Example of priced activity schedule, adapted from SBCSub/G

Mechanical services

A	Heating installation	235,000.00
B	Hot water installation	6,000.00
C	Cold water installation including rising main and tank	14,000.00
D	Gas installation	2,500.00
E	Mechanical ventilation	25,000.00
F	Controls and wiring	9,000.00
G	Testing and commissioning	3,000.00
H	As-installed record drawings	1,500.00
J	Operation and maintenance manuals	1,000.00
		£297,000.00

Electrical services

A	Distribution boards, switchgear and sub mains cabling	28,000.00
B	Power installation	72,000.00
C	Lighting installation including fittings	92,000.00
D	Fire alarm system	14,500.00
E	Security system	6,000.00
F	Earthing and bonding	5,000.00
G	Testing and certification	2,500.00
H	As-installed record drawings	1,250.00
J	Operation and maintenance manuals	750.00
		£222,000.00

References

Publications

Aeberli, P. *Focus on Construction Contract Formation*, RIBA Publishing, London (2003)

Construction Industry Council. *Building Information Model (BIM) Protocol* (CIC/BIM Pro), CIC, London (2013)

Fletcher, F. and Satchwell, H. *Briefing: A Practical Guide to RIBA Plan of Work 2013 Stages 7, 0 and 1*, RIBA Publishing, London (2015)

Furst, S. and Ramsay, V. *Keating on Construction Contracts*, 10th edn, Sweet & Maxwell, London (2016)

JCT. *JCT Standard Building Contract Guide* (SBC/G), Sweet & Maxwell, London (2016)

JCT. *JCT Standard Building Sub-Contract Guide* (SBCSub/G), Sweet & Maxwell, London (2016)

JCT/BDP. *The JCT Guide to the Use of Performance Specifications*, RIBA Publishing, London (2001)

Cases

Alexander v *Mercouris* [1979] 1 WLR 1270	3.17
Alfred McAlpine Capital Projects Ltd v *Tilebox Ltd* [2005] BLR 271	4.58
Alfred McAlpine Homes North Ltd v *Property and Land Contractors Ltd* (1995) 76 BLR 59	6.37
Archivent Sales & Developments Ltd v *Strathclyde Regional Council* (1984) 27 BLR 98 (Court of Session, Outer House)	5.59
B R Cantrell (2) E P Cantrell v *Wright and Fuller Ltd* [2003] BLR 412	7.44
Balfour Beatty Building Ltd v *Chestermont Properties Ltd* (1993) 62 BLR 1	4.33
Bath and North East Somerset District Council v *Mowlem plc* [2004] BLR 153 (CA)	5.51
BFI Group of Companies Ltd v *DCB Integration Systems Ltd* [1987] CILL 348	4.53, 4.58
Cavendish Square Holdings v *El Makdessi* and *ParkingEye Limited* v *Beavis* Supreme Court 2015	4.58
City Inn Ltd v *Shepherd Construction Ltd* [2008] CILL 2537 Outer House Court of Session	4.38
City of Westminster v *Jarvis & Sons Ltd* (1970) 7 BLR 64 (HL)	4.50
Colbart Ltd v *H Kumar* (1992) 59 BLR 89	7.50
Construction Partnership UK Ltd v *Leek Developments Ltd* [2006] CILL 2357 TCC	9.9
Co-operative Insurance Society v *Henry Boot Scotland and others* (2002) 84 Con LR 164	3.8
Croudace Ltd v *The London Borough of Lambeth* (1986) 33 BLR 20 (CA)	6.32
Crown Estate Commissioners v *John Mowlem & Co. Ltd* (1994) 70 BLR 1 (CA)	7.50
Dawber Williamson Roofing Ltd v *Humberside County Council* (1979) 14 BLR 70	5.62
Department of Environment for Northern Ireland v *Farrans (Construction) Ltd* (1982) 19 BLR 1 (NI)	4.65
Dhamija v *Sunningdale Joineries Ltd and others* [2010] EWHC 2396 (TCC)	7.17, 7.18
F G Minter Ltd v *Welsh Health Technical Services Organisation* (1980) 13 BLR 1 (CA)	6.37
Ferson Contractors Ltd v *Levolux A T Ltd* [2003] BLR 118	9.23
Galliford Try Building Ltd v *Estura Ltd* [2015] EWHC 412 (TCC)	7.35
Glenlion Construction Ltd v *The Guinness Trust* (1987) 39 BLR 89	4.14
Gloucestershire County Council v *Richardson* [1969] 1 AC 480 (CA)	3.27
Greater London Council v *Cleveland Bridge and Engineering Co.* (1986) 34 BLR 50 (CA)	4.10
H Fairweather & Co. Ltd v *London Borough of Wandsworth* (1987) 39 BLR 106	4.37, 6.40
H W Nevill (Sunblest) Ltd v *William Press & Son Ltd* (1981) 20 BLR 78	4.50

Wates Construction (London) Ltd v *Franthom Property Ltd* (1991) 53 BLR 23 (CA)   7.26
Wates Construction (South) Ltd v *Bredero Fleet Ltd* (1993) 63 BLR 128   6.21
West Faulkner Associates v *London Borough of Newham* (1992) 61 BLR 81   9.14
Whittal Builders Co. Ltd v *Chester-le-Street District Council* (1987) 40 BLR 82   4.4

Legislation

Statutes

Arbitration Act 1979   7.50
Arbitration Act 1996   10.21, 10.22, 10.27
Consumer Rights Act 2015   3.13, 4.2
Contracts (Rights of Third Parties) Act 1999   2.48–2.51
Defective Premises Act 1972   3.17
Employers' Liability (Compulsory Insurance) Act 1969   8.5
Freedom of Information Act 2000   1.17
Housing Grants, Construction and Regeneration Act 1996   1.13, 1.17, 2.28, 2.29, 7.4, 7.34, 7.35, 7.37, 9.10, 9.23, 10.8, 10.17, 10.18
Insolvency Act 1986   9.15
Law of Property Act 1925   2.45
Local Democracy, Economic Development and Construction Act 2009   1.13, 2.28, 7.4, 7.18, 7.22
Occupiers' Liability Acts 1957 and 1984   4.51
Sale of Goods Act 1979   3.13, 5.59, 5.60, 5.62
Supply of Goods and Services Act 1982   3.14, 4.2
Unfair Contract Terms Act 1977   3.13

Statutory instruments

Construction (Design and Management) Regulations   1.25, 1.26, 2.24–2.26, 3.32–3.35, 4.49, 5.8, 5.10, 5.20, 5.39, 5.56, 9.6, 9.11, 9.12
Late Payment of Commercial Debts Regulations 2013   1.17
Public Contracts Regulations 2015   1.17, 1.26, 9.6, 9.7, 9.32

Construction Industry Model Arbitration Rules   2.3, 10.22–10.25, 10.33
Insolvency Rules 2016   9.15

Construction Supply Chain Payment Charter 2014   1.17
Fair Payment Charter   1.17

Clause Index *by paragraph*

Subject Index *by paragraph*

A SINGLE STAR

A SINGLE STAR

AN ANTHOLOGY OF CHRISTMAS POETRY

COMPILED AND ARRANGED BY

David Davis

ILLUSTRATED BY

Margery Gill

THE BODLEY HEAD

LONDON · SYDNEY · TORONTO

To B.
with all my love

Printed and bound in Great Britain for
The Bodley Head Ltd
9 Bow Street, London WC2E 7AL
by BAS Printers Ltd, Over Wallop, Hampshire
Set in Monophoto Baskerville
First published 1973

Contents

Contents

Contents

Introduction

As all anthologists must do, I have tried, in this collection of Christmas poetry, to avoid including the most hackneyed and most obvious. Some of the old favourites are there, of course, but not many, because they can be found in so many other places. And I have tried to make the mixture not quite as before, but a combination of old (some of it very old) and new. Our old friend 'Anon.' (how he differs from 'Trad.' I am never quite sure) plays a big part, but so do poets of our own time, and particularly those of the inter-war and post-war period. It was a great pleasure to find how many of these there were. I once had an aunt who told me she disliked 'modern' music. When I asked her to be more precise, she said, 'Well, dear, Brahms, perhaps?' By that definition I suppose modern poetry might begin with Christina Rossetti! Though what I had in mind were poets of my own time and generation, and that covers not only the earlier part of the present century, but also the poetry of the fifties, sixties and seventies. Some names will occur again and again, and I make no apology for it. They are those for whom children and Christmas were always living realities: I think particularly of Walter de la Mare, and Eleanor Farjeon. But what splendid treasures there are in some of the other drawers. It is difficult, and perhaps a little invidious, to particularise. I have included nothing that did not light a small lamp of happiness inside me when I came across it. But I find deeply moving such things as James Kirkup's 'The Eve of Christmas': Ian Serraillier's 'The Mayor and the Simpleton': and Leonard Clark's two poems: 'Singing in the Streets', and 'Bells ringing'. 'The Computer's First Christmas Card' always makes me laugh. And I do feel, looking back on some, though not by any means all, of my own childhood Christmases, a certain wry sympathy with Elizabeth Jennings' poem which I have called 'Afterthought'; and fourteen-year-old Thomas Boyle's 'Christmas to me was Snow'. Even the best of Christmases sometimes have their snags.

The plan of the anthology is based partly on time, partly on the Gospel accounts of the birth of Jesus Christ. (Are we nowadays in danger of forgetting just what it is we are celebrating? I sometimes wonder.) It begins with the wonderful season of Anticipation, which I have associated, as we do in the northern hemisphere, in the best Christmas-card tradition, even if in defiance of reality, with winter and snow. Then, what for me has always been the most magical moment of all, the rapt waiting-time of Christmas Eve, leading on to the very Key of the Kingdom: Bethlehem, and the Stable, with the homage of the Shepherds. Christmas Day itself I have divided, like Julius Caesar, into three parts: morning, afternoon, and night; the afternoon, ('Make We Merry'), being given over to feasting and revelry, riddles, games, and the telling of tales. After the quiet farewell of Christmas Night, we look forward to Epiphany and After.

A word about the titles at the head of each poem: sometimes they are my own invention, intended to show the place they had in my own mind when I was choosing them. Occasionally I have kept the original title, when there was one. Often, it seemed better simply to quote the opening line.

Finally, who is it all aimed at? Children, I think, as far as possible: either to read to themselves, or to be read to. Some of it may be above their heads, but not all. One can never tell.

David Davis

I
Prologue

THE CHILDREN'S CAROL

Here we come again, again, and here we come again,
Christmas is a single pearl swinging on a chain,
Christmas is a single flower in a barren wood,
Christmas is a single sail on the salty flood,
Christmas is a single star in the empty sky,
Christmas is a single song sung for charity.
Here we come again, again, to sing to you again,
Give a single penny that we may not sing in vain.

Eleanor Farjeon

2
'Christmas Almost Come'

SNOW'S FALL'N DEEP

Now all the roads to London Town
Are windy-white with snow;
There's shouting and cursing,
And snortings to and fro;
But when night hangs her hundred lamps,
And the snickering frost-fires creep,
Then still, O; dale and hill, O;
Snow's fall'n deep.
Then still, O; dale and hill, O;
Snow's fall'n deep.

'Christmas Almost Come'

The carter cracks his leathery whip;
The ostler shouts Gee-whoa;
The farm dog grunts and sniffs and snuffs;
Bleat sheep; and cattle blow;
Soon Moll and Nan in dream are laid;
And snoring Dick's asleep;
Then still, O; dale and hill, O;
Snow's fall'n deep.
Then still, O; dale and hill, O;
Snow's fall'n deep.

Walter de la Mare

'Christmas Almost Come'

MERRY CHRISTMAS

Christmas comes! He comes, he comes,
Ushered with a rain of plums;
Hollies in the window greet him;
Schools come driving post to meet him,
Gifts precede him, bells proclaim him,
Every mouth delights to name him;
Wet, and cold, and wind, and dark,
Make him but the warmer mark;
And yet he comes not one-embodied,
Universal's the blithe godhead,
And in every festal house
Presence hath ubiquitous.
Curtains, those snug room-enfolders,
Hang upon his million shoulders,
And he has a million eyes
Of fire, and eats a million pies,
And is very merry and wise;
Very wise and very merry,
And loves a kiss beneath the berry.

from *Christmas*, Leigh Hunt

'Christmas Almost Come'

CHRISTMAS IS COMING

Christmas is coming,
   The geese are getting fat,
Please to put a penny
   In the old man's hat.
If you haven't got a penny,
   A ha'penny will do;
If you haven't got a ha'penny,
   Then God bless you!

St Thomas's Day is past and gone,
And Christmas almost come,
   Maidens arise,
   And make your pies,
And save young Bobby some.

Christmas comes but once a year,
And when it comes it brings good cheer,
A pocket full of money, and a cellar full of beer.

   Anon.

'Christmas Almost Come'

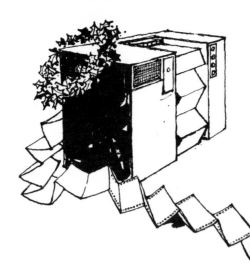

THE COMPUTER'S FIRST
CHRISTMAS CARD

jollymerry
hollyberry
jollyberry
merryholly
happyjolly
jollyjelly
jellybelly
bellymerry
hollyheppy
jollyMolly
marryJerry
merryHarry
hoppyBarry
heppyJarry
boppyheppy
berryjorry
jorryjolly
moppyjelly
Mollymerry

'Christmas Almost Come'

Jerryjolly
bellyboppy
jorryhoppy
hollymoppy
Barrymerry
Jarryhappy
happyboppy
boppyjolly
jollymerry
merrymerry
merrymerry
merryChris
ammerryasa
Chrismerry
asMERRYCHR
YSANTHEMUM

Edwin Morgan

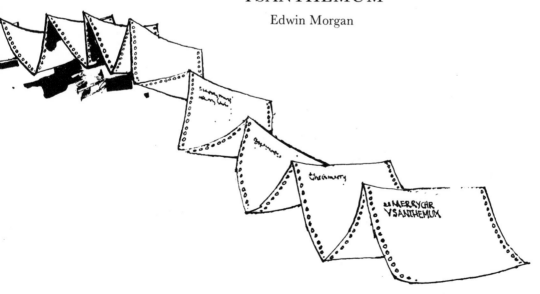

'Christmas Almost Come'

WELCOME YULE

Now, thrice welcome Christmas,
  Which brings us good cheer,
Minced pies and plum porridge,
  Good ale and strong beer;
With pig, goose, and capon,
  The best that can be,
So well doth the weather
  And our stomachs agree.

Observe how the chimneys
  Do smoke all about,
The cooks are providing
  For dinner no doubt;
But those on whose tables
  No victuals appear,
O may they keep Lent
  All the rest of the year!

With holly and ivy
  So green and so gay,
We deck up our houses
  As fresh as the day.
With bays and rosemary,
  And laurel complete;
And everyone now
  Is a king in conceit.

George Wither

(18)

'Christmas Almost Come'

WINTER

Green Mistletoe!
Oh, I remember now
A dell of snow,
Frost on the bough;
None there but I:
Snow, snow, and a wintry sky.

None there but I,
And footprints one by one,
Zigzaggedly,
Where I had run;
Where shrill and powdery
A robin sat in the tree.

And he whistled sweet;
And I in the crusted snow
With snow-clubbed feet
Jigged to and fro,
Till, from the day,
The rose-light ebbed away.

'Christmas Almost Come'

And the robin flew
Into the air, the air,
The white mist through;
And small and rare
The night-frost fell
Into the calm and misty dell.

And the dusk gathered low,
And the silver moon and stars
On the frozen snow
Drew taper bars,
Kindled winking fires
In the hooded briers.

And the sprawling Bear
Growled deep in the sky;
And Orion's hair
Streamed sparkling by:
But the North sighed low:
'*Snow, snow, more snow!*'

Walter de la Mare

'Christmas Almost Come'

THE ENDING OF THE YEAR

When trees did show no leaves,
  And grass no daisies had,
And fields had lost their sheaves,
  And streams in ice were clad,
And day of light was shorn,
  And wind had got a spear,
Jesus Christ was born
  In the ending of the year.

Like green leaves when they grow,
  He shall for comfort be;
Like life in streams shall flow,
  For running water He;
He shall raise hope like corn
  For barren fields to bear,
And therefore He was born
  In the ending of the year.

Like daisies to the grass,
  His innocence He'll bring;
In keenest winds that pass
  His flowering love shall spring;
The rising of the morn
  At midnight shall appear,
Whenever Christ is born
  In the ending of the year.

Eleanor Farjeon

3
Christmas Eve

ON CHRISTMAS EVE

On Christmas Eve I turned the spit,
I burnt my fingers, I feel it yet;
The little cock sparrow flew over the table,
The pot began to play with the ladle.
Anon.

Christmas Eve

SINGING IN THE STREETS

I had almost forgotten the singing in the streets,
Snow piled up by the houses, drifting
Underneath the door into the warm room,
Firelight, lamplight, the little lame cat
Dreaming in soft sleep on the hearth, mother dozing,
Waiting for Christmas to come, the boys and me
Trudging over blanket fields waving lanterns to the sky.
I had almost forgotten the smell, the feel of it all,
The coming back home, with girls laughing like stars,
Their cheeks, holly berries, me kissing one,
Silent-tongued, soberly, by the long church wall;
Then back to the kitchen table, supper on the white cloth,
Cheese, bread, the home-made wine:
Symbols of the Night's joy, a holy feast.
And I wonder now, years gone, mother gone,
The boys and girls scattered, drifted away with the snow-
    flakes,
Lamplight done, firelight over,
If the sounds of our singing in the streets are still there,
Those old times, still praising:
And now, a life-time of Decembers away from it all,
A branch of remembering holly spears my cheeks,
And I think it may be so;
Yes, I believe it may be so.

Leonard Clark

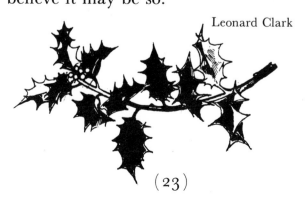

(23)

Christmas Eve

CAROL OF THE FIELD MICE

Villagers all, this frosty tide,
Let your doors swing open wide,
Though wind may follow, and snow beside,
Yet draw us in by your fire to bide,
   Joy shall be yours in the morning!

Here we stand in the cold and the sleet,
Blowing fingers and stamping feet,
Come from far away you to greet—
You by the fire and we in the street—
   Bidding you joy in the morning!

For ere one half of the night was gone,
Sudden a star has led us on,
Raining bliss and benison—
Bliss tomorrow and more anon,
   Joy for every morning!

Goodman Joseph toiled through the snow—
Saw the star o'er a stable low;
Mary she might not further go—
Welcome thatch, and litter below!
   Joy was hers in the morning!

And then they heard the angels tell
'Who were the first to cry Nowell?
Animals all, as it befell,
In the stable where they did dwell!
   Joy shall be theirs in the morning!'

Kenneth Grahame

Christmas Eve

A CAROL

Our Lord who did the Ox command
    To kneel to Judah's King,
He binds His frost upon the land
    To ripen it for Spring—
To ripen it for Spring, good sirs,
    According to His Word;
Which well must be as ye can see—
    And who shall judge the Lord?

When we poor fenmen skate the ice
    Or shiver on the wold,
We hear the cry of a single tree
    That breaks her heart in the cold—
That breaks her heart in the cold, good sirs,
    And rendeth by the board;
Which well must be as ye can see—
    And who shall judge the Lord?

Christmas Eve

Her wood is crazed and little worth
    Excepting as to burn,
That we may warm and make our mirth
    Until the Spring return—
Until the Spring return, good sirs,
    When people walk abroad;
Which well must be as ye can see—
    And who shall judge the Lord?

God bless the master of this house,
    And all who sleep therein!
And guard the fens from pirate folk,
    And keep us all from sin,
To walk in honesty, good sirs,
    Of thought and deed and word!
Which shall befriend our latter end . . .
    And who shall judge the Lord?

                Rudyard Kipling

EDDI'S SERVICE
(A.D. 687)

Eddi, priest of St Wilfrid
    In the chapel at Manhood End,
Ordered a midnight service
    For such as cared to attend.

But the Saxons were keeping Christmas,
    And the night was stormy as well.
Nobody came to service,
    Though Eddi rang the bell.

Christmas Eve

'Wicked weather for walking,'
    Said Eddi of Manhood End.
'But I must go on with the service
    For such as care to attend.'

The altar-lamps were lighted,—
    An old marsh-donkey came,
Bold as a guest invited,
    And stared at the guttering flame.

The storm beat on at the windows,
    The water splashed on the floor,
And a wet, yoke-weary bullock
    Pushed in through the open door.

'How do I know what is greatest,
    How do I know what is least?
That is my Father's business,'
    Said Eddi, Wilfrid's priest.

'But—three are gathered together—
    Listen to me and attend.
I bring good news, my brethren!'
    Said Eddi, of Manhood End.

And he told the Ox of a manger,
    And a stall in Bethlehem,
And he spoke to the Ass of a Rider
    That rode to Jerusalem.

Christmas Eve

They steamed and dripped in the chancel,
　　They listened and never stirred,
While, just as though they were Bishops,
　　Eddi preached them The Word.

Till the gale blew off on the marshes
　　And the windows showed the day,
And the Ox and the Ass together
　　Wheeled and clattered away.

And when the Saxons mocked him,
　　Said Eddi of Manhood End,
'I dare not shut His chapel
　　On such as care to attend.'

Rudyard Kipling

THE EVE OF CHRISTMAS

It was the evening before the night
That Jesus turned from dark to light.

Joseph was walking round and round,
And yet he moved not on the ground.

He looked into the heavens, and saw
The pole stood silent, star on star.

He looked into the forest: there
The leaves hung dead upon the air.

He looked into the sea, and found
It frozen, and the lively fishes bound.

Christmas Eve

And in the sky, the birds that sang
Not in feathered clouds did hang.

Said Joseph: 'What is this silence all?'
An angel spoke: 'It is no thrall,

But is a sign of great delight:
The Prince of Love is born this night.'

And Joseph said: 'Where may I find
This wonder?'—'He is all mankind,

Look, he is both farthest, nearest,
Highest and lowest, of all men the dearest.'

Then Joseph moved, and found the stars
Moved with him, and the evergreen airs,

The birds went flying, and the main
Flowed with its fishes once again.

And everywhere they went, they cried:
'Love lives, when all had died!'

In Excelsis Gloria!
James Kirkup

Christmas Eve

BELLS RINGING

I heard bells ringing
Suddenly all together, one wild, intricate figure,
A mixture of wonder and praise
Climbing the winter-winged air in December.
Norwich, Gloucester, Salisbury, combined with York
To shake Worcester and Paul's into the old discovery
Made frost-fresh again.
I heard these rocketing and wound-remembering chimes
Running their blessed counterpoint
Round the mazes of my mind,
And felt their message brimming over with love,
Watering my cold heart,
Until, as over all England hundreds of towers trembled
Beneath the force of Christmas rolling out,
I knew, as shepherds and magi knew,
That all sounds had been turned into one sound,
And a single golden bell,
Repeating, as knees bowed, the name EMMANUEL.

Leonard Clark

4
Bethlehem

O LITTLE TOWN

O little town of Bethlehem
   How still we see thee lie!
Above thy deep and dreamless sleep
   The silent stars go by.
Yet in thy dark streets shineth
   The everlasting light;
The hopes and fears of all the years
   Are met in thee tonight.

O morning stars, together
   Proclaim the holy birth,
And praises sing to God the King,
   And peace to men on earth;
For Christ is born of Mary;
   And, gathered all above,
While mortals sleep, the angels keep
   Their watch of wondering love.

Bethlehem

How silently, how silently,
  The wondrous gift is given!
So God imparts to human hearts
  The blessings of his heaven.
No ear may hear his coming;
  But in this world of sin,
Where meek souls will receive him, still
  The dear Christ enters in.

Where children pure and happy
  Pray to the blessèd Child,
Where misery cries out to thee,
  Son of the mother mild:
Where charity stands watching
  And faith holds wide the door,
The dark night wakes, the glory breaks,
  And Christmas comes once more.

O holy Child of Bethlehem,
  Descend to us, we pray;
Cast out our sin, and enter in,
  Be born in us today.
We hear the Christmas Angels
  The great glad tidings tell:
O come to us, abide with us,
  Our Lord Emmanuel.

Bishop Phillips Brooks

Bethlehem

HOW FAR TO BETHLEHEM?

How far is it to Bethlehem?
Not very far.
Shall we find the stable-room
Lit by a star?

Can we see the little child,
Is He within?
If we lift the wooden latch
May we go in?

May we stroke the creatures there,
Ox, ass and sheep?
May we peep like them and see
Jesus asleep?

If we touch His tiny hand
Will He awake?
Will He know we've come so far
Just for His sake?

Great kings have precious gifts,
And we have naught;
Little smiles and little tears
Are all we brought.

For all weary children
Mary must weep.
Here on His bed of straw
Sleep, children, sleep.

God in His mother's arms,
Babes in the byre
Sleep as they sleep who find
Their heart's desire.

Frances Chesterton

Bethlehem

A SINGING IN THE AIR

A snowy field! A stable piled
With straw! A donkey's sleepy pow!
A Mother beaming on a Child!
A Manger, and a munching cow!
—These we all remember now—
And airy voices, heard afar!
And three Magicians, and a Star!

Two thousand times of snow declare
That on the Christmas of the year
There is a singing in the air;
And all who listen for it hear
A fairy chime, a seraph strain,
Telling He is born again,
—That all we love is born again.

from *Christmas at Freelands*,
James Stephens

AN ODE ON THE BIRTH OF OUR SAVIOUR

In numbers, and but these few,
I sing thy birth, O Jesu!
Thou pretty baby, born here,
With superabundant scorn here,
Who for thy princely port here,
   Hadst for thy place
   Of birth, a base
Out-stable for thy court here.

Bethlehem

Instead of neat enclosures
Of interwoven osiers;
Instead of fragrant posies
Of daffodils, and roses:
Thy cradle, kingly stranger,
　　As gospel tells,
　　Was nothing else
But here a homely manger.

But we with silks, (not crewels)
With sundry precious jewels,
And lily-work will dress thee;
And, as we dispossess thee
Of clouts, we'll make a chamber,
　　Sweet babe, for thee,
　　Of ivory,
And plastered round with amber.

The Jews, they did disdain thee,
But we will entertain thee
With glories to await here
Upon thy princely state here,
And, more for love than pity,
　　From year to year
　　We'll make thee here
A free-born of our city.

Robert Herrick

Bethlehem

BEHOLD A SIMPLE TENDER BABE

Behold, a simple tender babe
    In freezing winter night
In homely manger trembling lies:
    Alas, a piteous sight!

The inns are full; no man will yield
    This little pilgrim bed;
But forced he is with simple beasts
    In crib to shroud his head.

Despise him not for lying there;
    First what he is inquire:
An orient pearl is often found
    In depth of dirty mire.

Weigh not his crib, his wooden dish,
    Nor beasts that by him feed;
Weigh not his mother's poor attire,
    Nor Joseph's simple weed.

This stable is a prince's court,
    This crib his chair of state,
The beasts are parcel of his pomp,
    The wooden dish his plate;

The persons in that poor attire
    His royal liveries wear;
The Prince himself is come from heaven,
    This pomp is prizèd there.

Bethlehem

With joy approach, O Christian wight,
  Do homage to thy King;
And highly praise his humble pomp,
  Which he from heaven doth bring.

Robert Southwell

LUTE-BOOK LULLABY

Sweet was the song the Virgin sang,
  When she to Bethlehem Judah came
And was delivered of a son,
  That blessèd Jesus hath to name:
    Lulla, lulla, lulla, lulla-by:
    Lulla, lulla, lulla, lulla-by.

'Sweet babe,' sang she, 'my Son,
  And eke a Saviour born,
Who hast vouchsafèd from on high
  To visit us that were forlorn:
    Lalula, lalula, lalula-by.'
'Sweet babe,' sang she,
And rocked him sweetly on her knee.

Anon.

(37)

Bethlehem

A CHRISTMAS CAROL

The Christ-child lay on Mary's lap,
  His hair was like a light.
(O weary, weary were the world,
  But here is all aright.)

The Christ-child lay on Mary's breast,
  His hair was like a star.
(O stern and cunning are the Kings,
  But here the true hearts are.)

The Christ-child lay on Mary's heart,
  His hair was like a fire.
(O weary, weary, is the world,
  But here the world's desire.)

The Christ-child stood at Mary's knee,
  His hair was like a crown,
And all the flowers looked up at him,
  And all the stars looked down.

G. K. Chesterton

Bethlehem

CAROL

Sing, happy child, Noël, Noël,
Bright shines Orion's sword
Where every star stands sentinel
And watchful of their Lord.

Sweetly the carol singers speak,
They fill the firelit hall,
Singing of Mary, fair and meek,
And Jesus in the stall.

Hark, happy child, to what they say,
Lock in your heart their song
Lest you should lose it on the way
When every road seems long.

You will recall the spicèd scent
Of leaves where no winds stir,
When gold and frankincense are spent,
And nothing's left but myrrh.

                    Eiluned Lewis

LULLABY

Sleep, my baby, the night is coming soon.
Sleep, my baby, the day has broken down.

Sleep now: let silence come, let the shadows form
A castle of strength for you, a fortress of calm.

You are so small, sleep will come with ease.
Hush now, be still now, join the silences.

                    Elizabeth Jennings

Bethlehem

THE SHEPHERDS

From far away we come to you,
The snow in the street and the wind on the door,
To tell of great tidings strange and true.
Minstrels and maids stand forth on the floor:
From far away we come to you,
To tell of great tidings strange and true.

For as we wandered far and wide,
The snow in the street and the wind on the door,
What hap do you deem there should us betide?
Minstrels and maids stand forth on the floor:
From far away we come to you,
To tell of great tidings strange and true.

Under a bent when the night was deep,
The snow in the street and the wind on the door,
There lay three shepherds tending their sheep:
Minstrels and maids stand forth on the floor:
From far away we come to you,
To tell of great tidings strange and true.

'O ye shepherds, what have ye seen,
The snow in the street and the wind on the door,
To slay your sorrow and heal your teen?'
Minstrels and maids stand forth on the floor:
From far away we come to you,
To tell of great tidings strange and true.

(40)

Bethlehem

'In an ox-stall this night we saw
The snow in the street and the wind on the door,
A babe and a maid without a flaw:
Minstrels and maids stand forth on the floor:
From far away we come to you,
To tell of great tidings strange and true.

'There was an old man there beside:
The snow in the street and the wind on the door,
His hair was white and his hood was wide:
Minstrels and maids stand forth on the floor:
From far away we come to you,
To tell of great tidings strange and true.

'And as we gazed this thing upon,
The snow in the street and the wind on the door,
Those twain knelt down to the little one.'
Minstrels and maids stand forth on the floor:
From far away we come to you,
To tell of great tidings strange and true.

'And a marvellous song we straight did hear,
The snow in the street and the wind on the door,
That slew our sorrow and healed our care.'
Minstrels and maids stand forth on the floor:
From far away we come to you,
To tell of great tidings strange and true.

News of a fair and a marvellous thing,
The snow in the street and the wind on the door,
Nowell, nowell, nowell, we sing!
Minstrels and maids stand forth on the floor:
From far away we come to you,
To tell of great tidings strange and true.

William Morris

Bethlehem

THE MAYOR AND THE SIMPLETON

They followed the Star to Bethlehem—
Boolo the baker, Barleycorn the farmer,
old Darby and Joan, a small boy Peter, and
a simpleton whose name was Innocent.
Over the snowfields and the frozen rutted lanes
they followed the Star to Bethlehem.

Innocent stood at the stable door
and watched them enter. A flower
stuck out of his yellow hair; his mouth gaped open
like a drawer that wouldn't shut.
He beamed upon the child where he lay
among the oxen, in swaddling clothes in the hay,
his blue eyes shining steady as the Star overhead;
beside him old Joseph and
Mary his mother, smiling.

                          Innocent was delighted.

Bethlehem

They brought gifts with them—Boolo, some fresh crusty
   loaves
(warm from the baking) which he laid
at the feet of the infant Jesus, kneeling
in all humility.

                     Innocent was delighted.

Barleycorn brought two baskets—one with a dozen eggs,
the other with two chickens—which he laid
at the feet of the infant Jesus, kneeling
in all humility.

                     Innocent was delighted.

Bethlehem

Darby and Joan brought apples and pears from their garden,
wrapped in her apron and stuffed
in the pockets of his trousers; the little boy
a pot of geraniums—he had grown them himself.
And they laid them
at the feet of the infant Jesus, kneeling
in all humility.

      Innocent was delighted.

The mayor rolled up in his coach with a jingle of bells
and a great to-do. He stepped out with a flourish
and fell flat on his face in the snow. His footmen
picked him up and opened his splendid
crimson umbrella. Then he strutted to the door,
while the white flakes floated down
and covered it with spots. He was proud of his umbrella
and didn't mean to give it away.

Bethlehem

Shaking the snow off on to the stable floor,
the mayor peered down at the child where he lay
among the oxen, in swaddling clothes in the hay,
his blue eyes shining steady as the Star overhead,
beside him old Joseph and
Mary his mother, smiling.

                         Innocent was puzzled.

And the mayor said: 'On this important occasion
each must take a share in the general thanksgiving.
Hence the humble gifts—the very humble gifts—
which I see before me. My own contribution
is something special—a speech. I made it up myself and I'm
   sure
you'll all like it. Ahem. Pray silence for the mayor.'

Bethlehem

'Moo, moo,' said the oxen.
                          'My fellow citizens,
the happy event I refer to—in which we all rejoice—
has caused a considerable stir
in the parish—'
                '—in the whole world,' said a voice.

Who spoke? Could it be Innocent, always so shy,
timid as a butterfly, frightened
as a sparrow with a broken wing? Yes, it was he.
Now God had made him bold.

 'I fear I must start again,' said the mayor.
'My fellow citizens, in the name of the people of this parish
I am proud to welcome one
who promises so well—'
                  '—He is the Son of Heaven,'
said Innocent.

Bethlehem

The mayor took no notice.
'I prophesy a fine future for him,
almost—you might say—spectacular.
He'll do us all credit. At the same time I salute in particular
the child's mother, the poor woman who—'
    '—She is not poor but the richest, most radiant
of mothers.'
    'Simpleton, how dare you interrupt!'
snapped the mayor.
But God, who loves the humble, heard him not.
He made him listen, giving Innocent the words:
'Mr Mayor, you don't understand. This birth
is no local event. The child is Jesus,
King of kings and Lord of lords.
A stable is his place and poverty his dwelling-place—
yet he has come to save the world. No speech
is worthy of him—'
                'Tush!' said the mayor.

'I took a lot of trouble. It's a rare
and precious gift, my speech—and now
I can't get a word in edgeways.'
    'Rare and precious, did you say? Hear what the child
has brought to *us*—peace on earth, goodwill toward men.
O truly rare and precious gift!'
    'Peace on earth,' said the neighbours,
'goodwill toward men. O truly rare
and precious gift!' They knelt in humility,
in gratitude to the child who lay
among the oxen, in swaddling clothes in the hay,
his blue eyes shining steady as the Star overhead,
beside him old Joseph and
Mary his mother, smiling.

(47)

Bethlehem

The mayor was silent. God gave the simpleton
no more to say. Now
like a frightened bird
over the snowfields and the frozen rutted lanes
he fluttered away. Always, as before, a flower
stuck out of his yellow hair; his mouth gaped open
like a drawer that wouldn't shut.
He never spoke out like that again.

As for the mayor, he didn't finish his speech.
He called for his coach and drove off, frowning,
much troubled. For a little while
he thought of what the simpleton had said
But he soon forgot all about it, having
important business to attend to in town.

Ian Serraillier

Bethlehem

THE KEY OF THE KINGDOM

This is the key of the kingdom:
In that kingdom is a city,
In that city is a town,
In that town there is a street,
In that street there winds a lane,
In that lane there is a yard,
In that yard there is a house,
In that house there waits a room,
In that room there is a bed,
On that bed there is a basket.
   A basket of flowers.

Flowers in the basket,
Basket on the bed,
Bed in the chamber,
Chamber in the house,
House in the weedy yard,
Yard in the winding lane,
Lane in the broad street,
Street in the high town,
Town in the city,
City in the kingdom:
   This is the key of the kingdom.

Anon.

Bethlehem

THE INNKEEPER'S WIFE

I love this byre. Shadows are kindly here.
The light is flecked with travelling stars of dust.
So quiet it seems after the inn-clamour,
Scraping of fiddles and the stamping feet.
Only the cows, each in her patient box,
Turn their slow eyes, as we and the sunlight enter,
Their slowly rhythmic mouths.
                              'That is the stall,
Carpenter. You see it's too far gone
For patching or repatching. My husband made it,
And he's been gone these dozen years and more . . .'
Strange how this lifeless thing, degraded wood
Split from the tree and nailed and crucified
To make a wall, outlives the mastering hand
That struck it down, the warm firm hand
That touched my body with its wandering love.
'No, let the fire take them. Strip every board
And make a new beginning. Too many memories lurk
Like worms in this old wood. That piece you're holding—
That patch of grain with the giant's thumbprint—
I stared at it a full hour when he died:
Its grooves are down my mind. And that board there
Baring its knot-hole like a missing jig-saw—
I remember another hand along its rim.
No, not my husband's, and why I should remember
I cannot say. It was a night in winter.
Our house was full, tight-packed as salted herrings—
So full, they said, we had to hold our breaths
To close the door and shut the night-air out!

(50)

Bethlehem

And then two travellers came. They stood outside
Across the threshold, half in the ring of light
And half beyond it. I would have let them in
Despite the crowding—the woman was past her time—
But I'd no mind to argue with my husband,
The flagon in my hand and half the inn
Still clamouring for wine. But when trade slackened,
And all our guests had sung themselves to bed
Or told the floor their troubles, I came out here
Where he had lodged them. The man was standing
As you are now, his hand smoothing that board.—
He was a carpenter, I heard them say.
She rested on the straw, and on her arm
A child was lying. None of your creased-faced brats
Squalling their lungs out. Just lying there
As calm as a new-dropped calf—his eyes wide open,
And gazing round as if the world he saw
In the chaff-strewn light of the stable lantern
Was something beautiful and new and strange.
Ah well, he'll have learnt different now, I reckon,
Wherever he is. And why I should recall
A scene like that, when times I would remember
Have passed beyond reliving, I cannot think.
It's a trick you're served by old possessions:
They have their memories too—too many memories.
Well, I must go in. There are meals to serve.
Join us there, Carpenter, when you've had enough
Of cattle-company. The world is a sad place,
But wine and music blunt the truth of it.'

Clive Sansom

5
Christmas Morning

WELCOME TO HEAVEN'S KING

Welcome be Thou, Heaven's King,
Welcome, born in one morning,
Welcome, for Him we shall sing,
Welcome, Yule!

Anon.

Christmas Morning

LITTLE CHRIST JESUS

Now every Child that dwells on earth,
Stand up, stand up and sing:
The passing night has given birth
Unto the children's King.
Sing sweet as the flute,
Sing clear as the horn,
Sing joy of the Children,
Come Christmas the morn:
  Little Christ Jesus
  Our brother is born.

Now every star that dwells in sky,
Look down with shining eyes:
The night has dropped in passing by
A Star from Paradise.
Sing sweet as the flute,
Sing clear as the horn,
Sing joy of the Stars,
Come Christmas the morn:
  Little Christ Jesus
  Our brother is born.

Now every Beast that crops in field,
Breathe sweetly and adore:
The night has brought the richest yield
That ever the harvest bore.
Sing sweet as the flute,
Sing clear as the horn,
Sing joy of the Creatures,
Come Christmas the morn:
  Little Christ Jesus
  Our brother is born.

Christmas Morning

Now every Bird that flies in air,
Sing, raven, lark and dove:
The night has brooded on her lair
And fledged the Bird of love.
Sing sweet as the flute,
Sing clear as the horn,
Sing joy of the Birds,
Come Christmas the morn:
Little Christ Jesus
Our brother is born.

Now all the Angels of the Lord,
Rise up on Christmas Even:
The passing night will hear the Word
That is the voice of Heaven.
Sing sweet as the flute,
Sing clear as the horn,
Sing joy of the Angels,
Come Christmas the morn:
Little Christ Jesus
Our brother is born.

Eleanor Farjeon

IT WAS ON CHRISTMAS DAY

It was on Christmas Day,
And all in the morning,
Our Saviour was born,
And our heavenly King:
And was not this a joyful thing?
And sweet Jesus they called him by name.

Anon.

Christmas Morning

AS I SAT ON A SUNNY BANK

As I sat on a sunny bank
On Christmas day in the morning,
I saw three ships come sailing by
On Christmas day in the morning.
And who do you think were in those ships
But Joseph and his fair lady:
He did whistle and she did sing,
And all the bells on earth did ring
For joy our Saviour He was born
On Christmas day in the morning.

Anon.

Christmas Morning

BEFORE THE PALING OF THE STARS

Before the paling of the stars,
   Before the winter morn,
   Before the earliest cock-crow,
Jesus Christ was born:
     Born in a stable,
   Cradled in a manger,
In the world His hands had made
     Born a stranger.

Priest and King lay fast asleep
   In Jerusalem;
Young and old lay fast asleep
   In crowded Bethlehem;
Saint and Angel, ox and ass,
   Kept a watch together,
   Before the Christmas daybreak
   In the winter weather.

Jesus on His Mother's breast
   In the stable cold,
Spotless Lamb of God was He,
   Shepherd of the fold:
Let us kneel with Mary Maid,
   With Joseph bent and hoary,
With Saint and Angel, ox and ass,
   To hail the King of Glory.

Christina Rossetti

Christmas Morning

AFTERTHOUGHT

For weeks before it comes I feel excited, yet when it
At last arrives, things all go wrong:
My thoughts don't seem to fit.

I've planned what I'll give everyone and what they'll give
   to me,
And then on Christmas morning all
The presents seem to be

Useless and tarnished. I have dreamt that everything would
   come
To life—presents and people too.
Instead of that, I'm dumb,

And people say, 'How horrid! What a sulky little boy!'
And they are right. I *can't* seem pleased.
The lovely shining toy

I wanted so much when I saw it in a magazine
Seems pointless now. And Christmas too
No longer seems to mean

The hush, the star, the baby, people being kind again.
The bells are rung, sledges are drawn,
And peace on earth for men.

Elizabeth Jennings

(57)

6

'Make We Merry'

MAKE WE MERRY

Make we merry, both more and less,
For now is the time of Christemas.

Let no man come into this hall,
Nor groom, nor page, not yet marshall,
But that some sport he bring withal.

If that he say he cannot sing,
Some other sport then let him bring,
That it may please at this feasting.

If he say he naught can do,
Then, for my love, ask him no mo'
But to the stocks then let him go.

Make we merry, both more and less,
For now is the time of Christemas.

Anon.

'Make We Merry'

OUR JOYFULL'ST FEAST

So, now is come our joyfull'st feast;
　　Let every man be jolly.
Each room with ivy-leaves is dressed,
　　And every post with holly.
Though some churls at our mirth repine,
Round your foreheads garlands twine,
Drown sorrow in a cup of wine,
　　And let us all be merry.

Now, all our neighbours' chimneys smoke,
　　And Christmas blocks are burning;
Their ovens, they with baked meats choke,
　　And all their spits are turning.
Without the door let sorrow lie,
And if for cold it hap to die,
We'll bury it in a Christmas pie,
　　And evermore be merry.

from *A Christmas Carol*, George Wither

'Make We Merry'

WASSAIL!

Wassail, wassail, all over the town!
Our toast it is white, and our ale it is brown,
Our bowl it is made of the white maple tree;
With the wassailing bowl we'll drink to thee!

So here is to Cherry and to his right cheek,
Pray God send our master a good piece of beef,
And a good piece of beef that may we all see;
With the wassailing bowl we'll drink to thee!

And here is to Dobbin and to his right eye,
Pray God send our master a good Christmas pie,
And a good Christmas pie that may we all see;
With our wassailing bowl we'll drink to thee!

'Make We Merry'

So here is to Broad May and to her broad horn,
May God send our master a good crop of corn,
And a good crop of corn that may we all see;
With the wassailing bowl we'll drink to thee!

And here is to Fill-Pail and to her left ear,
Pray God send our master a happy New Year,
And a happy New Year as e'er he did see;
With our wassailing bowl we'll drink to thee!

And here is to Colly and to her long tail,
Pray God send our master he never may fail
A bowl of strong beer; I pray you draw near,
And our jolly wassail it's then you shall hear!

Come, butler, come fill us a bowl of the best,
Then we hope that your soul in heaven may rest:
But if you do draw us a bowl of the small,
Then down shall go butler, bowl and all.

Then here's to the maid in the lily-white smock,
Who tripped to the door and slipped back the lock!
Who tripped to the door and pulled back the pin,
For to let these jolly wassailers in!

Anon.

'Make We Merry'

SOME CHRISTMAS RIDDLES

Flour of England, fruit of Spain,
Met together in a shower of rain;
Put in a bag, tied round with a string;
If you tell me this riddle,
I'll give you a ring.
(Plum pudding)

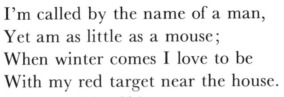

I'm called by the name of a man,
Yet am as little as a mouse;
When winter comes I love to be
With my red target near the house.
(A robin)

Highty, tighty, paradighty,
Clothed all in green,
The king could not read it,
No more could the queen;
They sent for the wise men
From out of the East,
Who said it had horns,
But was not a beast.
(A holly-leaf)

Anon.

'Make We Merry'

SNAPDRAGON

Here he comes with flaming bowl,
Don't he mean to take his toll,
  Snip! Snap! Dragon.
Take care you don't take too much,
Be not greedy in your clutch,
  Snip! Snap! Dragon.

With his blue and lapping tongue
Many of you will be stung,
  Snip! Snap! Dragon.
For he snaps at all that comes
Snatching at his feast of plums,
  Snip! Snap! Dragon.

But old Christmas makes him come,
Though he looks so fee! fo! fum!
  Snip! Snap! Dragon.
Don't 'ee fear him, but be bold,
Out he goes, his flames are cold,
  Snip! Snap! Dragon.
                    Anon.

'Make We Merry'

A SONG FOR ANYONE TO SING

There was a pig went out to dig,
On Chrisimas Day, Chrisimas Day,
There was a pig went out to dig
On Chrisimas Day in the morning.

There was a cow went out to plough,
On Chrisimas Day, Chrisimas Day.
There was a cow went out to plough
On Chrisimas Day in the morning.

There was a doe went out to hoe,
On Chrisimas Day, Chrisimas Day,
There was a doe went out to hoe
On Chrisimas Day in the morning.

There was a drake went out to rake,
On Chrisimas Day, Chrisimas Day,
There was a drake went out to rake
On Chrisimas Day in the morning.

'Make We Merry'

There was a sparrow went out to harrow,
On Chrisimas Day, Chrisimas Day,
There was a sparrow went out to harrow
On Chrisimas Day in the morning.

There was a minnow went out to winnow,
On Chrisimas Day, Chrisimas Day,
There was a minnow went out to winnow
On Chrisimas Day in the morning.

There was a sheep went out to reap,
On Chrisimas Day, Chrisimas Day,
There was a sheep went out to reap
On Chrisimas Day in the morning.

There was a crow went out to sow,
On Chrisimas Day, Chrisimas Day,
There was a crow went out to sow
On Chrisimas Day in the morning.

There was a row went out to mow,
On Chrisimas Day, Chrisimas Day,
There was a row went out to mow
On Chrisimas Day in the morning.

Anon.

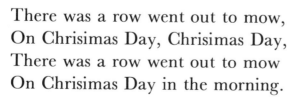

'Make We Merry'

GHOST STORY

Bring out the tall tales now that we told
by the fire as the gaslight bubbled like a diver.
Ghosts whooed like owls in the long nights
when I dared not look over my shoulder; animals
lurked in the cubbyhole under the stairs where the
gas meter ticked. And I remember that we went
singing carols once, when there wasn't the shaving
of a moon to light the flying streets. At the end
of a long road was a drive that led to a large
house, and we stumbled up the darkness of the drive
that night, each one of us afraid, each one holding
a stone in his hand in case, and all of us too brave
to say a word. The wind through the trees
made noises as of old and unpleasant and maybe
webfooted men wheezing in caves. We reached
the black bulk of the house.
'What shall we give them? Hark the Herald?'
'No,' Jack said, 'Good King Wenceslas.
I'll count three.'

'Make We Merry'

One, two, three, and we began to sing,
our voices high and seemingly distant in the
snow-felted darkness round the house that
was occupied by nobody we knew. We stood
close together, near the dark door.
'Good King Wenceslas looked out
On the Feast of Stephen . . .'
And then a small, dry voice, like the voice
of someone who has not spoken for a long time,
joined our singing: a small dry eggshell voice
from the other side of the door: a small dry voice
through the keyhole. And when we stopped running
we were outside *our* house; the front room was lovely:
balloons floated under the hot-water-bottle-gulping gas;
everything was good again and shone over the town.

'Perhaps it was a ghost,' Jim said.
'Perhaps it was trolls,' Dan said,
who was always reading.

'Let's go in and see if there's any jelly left,'
Jack said. And we did that.

Dylan Thomas

'Make We Merry'

CHRISTMAS TO ME WAS SNOW

Christmas to me
was snow
but it never snowed
it always rained
or was sunny.
Once it snowed
and that was Christmas.
But the turkey got burnt
and when you chewed it
Mum said 'Do you like it?'
and you said 'Yes'
and that was her Christmas.
Dad's was a cigar
or an ounce of St Bruno or new slippers.

Thomas Boyle (aged 14)

(68)

7
Christmas Night

THIS HOLY NIGHT

God bless your house this holy night,
　And all within it;
God bless the candle that you light
　To midnight's minute:
The board at which you break your bread,
　The cup you drink of:
And as you raise it, the unsaid
　Name that you think of:
The warming fire, the bed of rest,
　The ringing laughter:
These things, and all things else be blest
　From floor to rafter
This holy night, from dark to light,
　Even more than other;
And, if you have no house tonight,
　God bless you, brother.

Eleanor Farjeon

Christmas Night

HOLLY RED AND MISTLETOE WHITE

Holly red and mistletoe white,
The stars are shining with golden light,
Burning like candles this Holy Night,
Holly red and mistletoe white.

Mistletoe white and holly red,
The doors are shut and the children a-bed,
Fairies at foot and angels at head,
Mistletoe white and holly red.

Alison Uttley

Christmas Night

THE FIRST TREE IN THE GREENWOOD

Now the holly bears a berry as white as the milk,
And Mary bore Jesus, who was wrapped up in silk:
    And Mary bore Jesus Christ,
    Our Saviour for to be,
    And the first tree in the greenwood, it was the holly.

Now the holly bears a berry as green as the grass,
And Mary bore Jesus, who died on the cross:
    And Mary bore Jesus Christ,
    Our Saviour for to be,
    And the first tree in the greenwood, it was the holly.

Now the holly bears a berry as black as the coal,
And Mary bore Jesus, who died for us all:
    And Mary bore Jesus Christ,
    Our Saviour for to be,
    And the first tree in the greenwood, it was the holly.

Now the holly bears a berry, as blood is it red,
Then trust we our Saviour, who rose from the dead:
    And Mary bore Jesus Christ,
    Our Saviour for to be,
    And the first tree in the greenwood, it was the holly.

Anon.

Christmas Night

THE CHRISTMAS TREE

Put out the lights now!
Look at the Tree, the rough tree dazzled
In oriole plumes of flame,
Tinselled with twinkling frost fire, tasselled
With stars and moons—the same
That yesterday hid in the spinney and had no fame
Till we put out the lights now.

Hard are the nights now:
The fields at moonrise turn to agate,
Shadows are cold as jet;
In dyke and furrow, in copse and faggot
The frost's tooth is set;
And stars are the sparks whirled out by the north wind's fret
On the flinty nights now.

So feast your eyes now
On mimic star and moon-cold bauble:
Worlds may wither unseen,
But the Christmas Tree is a tree of fable,
A phoenix in evergreen,
And the world cannot change or chill what its mysteries mean
To your hearts and eyes now.

The vision dies now
Candle by candle: the tree that embraced it
Returns to its own kind,
To be earthed again and weather as best it
May the frost and the wind.
Children, it too had its hour—you will not mind
If it lives or dies now.

C. Day Lewis

(72)

Christmas Night

YULE LOG

Kindle the Christmas brand, and then
  Till sunset let it burn;
Which quenched, then lay it up again,
  Till Christmas next return.

Part must be kept wherein to tend
  The Christmas log next year,
And where 'tis safely kept, the fiend
  Can do no mischief there.

Robert Herrick

Christmas Night

LOVE CAME DOWN AT CHRISTMAS

Love came down at Christmas,
    Love all lovely, Love Divine;
Love was born at Christmas,
    Star and angels gave the sign.

Worship we the Godhead,
    Love Incarnate, Love Divine;
Worship we our Jesus:
    But wherewith for sacred sign?

Love shall be our token,
    Love be yours and love be mine,
Love to God and all men,
    Love for plea and gift and sign.

Christina Rossetti

8
Epiphany and After

ALL IN THE MORNING

It was on the Twelfth Day,
And all in the morning,
The Wise Men were led
To our heavenly King;
And was not this a joyful thing?
And sweet Jesus they called him by name.

It was on Twentieth Day,
And all in the morning,
The Wise Men returned
From our heavenly King;
And was not this a joyful thing?
And sweet Jesus they called him by name.

Anon.

Epiphany and After

THREE KINGS CAME RIDING

Three Kings came riding from far away,
  Melchior and Gaspar and Baltasar;
Three Wise Men out of the East were they,
And they travelled by night and they slept by day,
  For their guide was a beautiful, wonderful star.

The star was so beautiful, large, and clear,
  That all the other stars of the sky
Became a white mist in the atmosphere,
And by this they knew that the coming was near
  Of the Prince foretold in the prophecy.

Three caskets they bore on their saddle-bows,
  Three caskets of gold with golden keys;
Their robes were of crimson silk, with rows
Of bells and pomegranates and furbelows,
  Their turbans like blossoming almond-trees.

And so the Three Kings rode into the West,
  Through the dusk of night, over hill and dell,
And sometimes they nodded, with beard on breast,
And sometimes talked, as they paused to rest,
  With the people they met at some wayside well.

'Of the child that is born,' said Baltasar,
  'Good people, I pray you, tell us the news;
For we in the East have seen his star,
And have ridden fast, and have ridden far,
  To find and worship the King of the Jews.'

Epiphany and After

And the people answered, 'You ask in vain;
   We know of no king but Herod the Great!'
They thought the Wise Men were men insane,
As they spurred their horses across the plain,
   Like riders in haste, and who cannot wait.

And when they came to Jerusalem,
   Herod the Great, who had heard this thing,
Sent for the Wise Men and questioned them;
And said, 'Go down unto Bethlehem,
   And bring me tidings of this new king.'

So they rode away; and the star stood still,
   The only one in the grey of morn;
Yes, it stopped, it stood still of its own free will,
Right over Bethlehem on the hill,
   The city of David where Christ was born.

And the Three Kings rode through the gate and the guard,
   Through the silent street, till their horses turned
And neighed as they entered the great inn-yard;
But the windows were closed, and the doors were barred,
   And only a light in the stable burned.

And cradled there in the scented hay,
   In the air made sweet by the breath of kine,
The little child in the manger lay,
The child that would be King one day
   Of a kingdom not human but divine.

His mother, Mary of Nazareth,
   Sat watching beside his place of rest,
Watching the even flow of his breath,
For the joy of life and the terror of death
   Were mingled together in her breast.

Epiphany and After

They laid their offerings at his feet:
    The gold was their tribute to a King,
The frankincense, with its odour sweet,
Was for the Priest, the Paraclete,
    The myrrh for the body's burying.

And the mother wondered and bowed her head,
    And sat as still as a statue of stone;
Her heart was troubled yet comforted,
Remembering what the Angel had said,
    Of an endless reign and of David's throne.

Then the Kings rode out of the city gate,
    With a clatter of hoofs in proud array;
But they went not back to Herod the Great,
For they knew his malice and feared his hate,
    And returned to their homes by another way.

Henry Wadsworth Longfellow

THE STRANGERS

Dim-berried is the mistletoe
With globes of sheenless grey,
The holly mid ten thousand thorns
Smoulders its fires away;
And in the manger Jesu sleeps
    This Christmas Day.

Epiphany and After

Bull unto bull with hollow throat
Makes echo every hill,
Cold sheep in pastures thick with snow
The air with bleatings fill;
While of his mother's heart this Babe
　　Takes His sweet will.

All flowers and butterflies lie hid,
The blackbird and the thrush
Pipe but a little as they flit
Restless from bush to bush;
Even to the robin Gabriel hath
　　Cried softly, 'Hush!'

Now night's astir with burning stars
In darkness of the snow;
Burdened with frankincense and myrrh
And gold the Strangers go
Into a dusk where one dim lamp
　　Burns faintly, Lo!

No snowdrop yet its small head nods,
In winds of winter drear;
No lark at casement in the sky
Sings matins shrill and clear;
Yet in this frozen mirk the Dawn
　　Breathes, Spring is here!

Walter de la Mare

Epiphany and After

IN A FAR LAND UPON A DAY

In a far land upon a day,
Where never snow did fall,
Three Kings went riding on the way
Bearing presents all.

And one wore red, and one wore gold,
And one was clad in green,
And one was young, and one was old,
And one was in between.

The middle one had human sense,
The young had loving eyes,
The old had much experience,
And all of them were wise.

Choosing no guide by eve and morn
But heaven's starry drifts,
They rode to find the Newly-Born
For whom they carried gifts.

Oh, far away in time they rode
Upon their wanderings,
And still in story goes abroad
The riding of the Kings:

So wise, that in their chosen hour,
As through the world they filed,
They sought not wealth or place or power,
But rode to find a child.

Eleanor Farjeon

Epiphany and After

JOURNEY OF THE MAGI

'A cold coming we had of it,
Just the worst time of the year
For a journey, and such a long journey:
The ways deep and the weather sharp,
The very dead of winter.'
And the camels galled, sore-footed, refractory,
Lying down in the melting snow.
There were times we regretted
The summer palaces on slopes, the terraces,
And the silken girls bringing sherbet.
Then the camel men cursing and grumbling
And running away, and wanting their liquor and women,
And the night-fires going out, and the lack of shelters,
And the cities hostile and the towns unfriendly
And the villages dirty and charging high prices:
A hard time we had of it.
At the end we preferred to travel all night,
Sleeping in snatches,
With the voices singing in our ears, saying
That this was all folly.

Epiphany and After

Then at dawn we came down to a temperate valley,
Wet, below the snow line, smelling of vegetation;
With a running stream and a water-mill beating the darkness,
And three trees on the low sky,
And an old white horse galloped away in the meadow.
Then we came to a tavern with vine-leaves over the lintel,
Six hands at an open door dicing for pieces of silver,
And feet kicking the empty wine-skins.
But there was no information, and so we continued
And arrived at evening, not a moment too soon
Finding the place: it was (you may say) satisfactory.

All this was a long time ago, I remember,
And I would do it again, but set down
This set down
This: were we led all that way for
Birth or Death? There was a Birth, certainly,
We had evidence and no doubt. I had seen birth and death,
But had thought they were different; this Birth was
Hard and bitter agony for us, like Death, our death.
We returned to our places, these Kingdoms,
But no longer at ease here, in the old dispensation,
With an alien people clutching their gods.
I should be glad of another death.

T. S. Eliot

(82)

Epiphany and After

INNOCENT'S SONG

Who's that knocking on the window,
Who's that standing at the door,
What are all those presents
Lying on the kitchen floor?

Who is the smiling stranger
With hair as white as gin,
What is he doing with the children
And who could have let him in?

Why has he rubies on his fingers,
A cold, cold crown on his head,
Why, when he caws his carol,
Does the salty snow run red?

Why does he ferry my fireside
As a spider on a thread,
His fingers made of fuses
And his tongue of gingerbread?

Why does the world before him
Melt in a million suns,
Why do his yellow, yearning eyes
Burn like saffron buns?

Watch where he comes walking
Out of the Christmas flame,
Dancing, double-talking:

Herod is his name.

Charles Causley

Epiphany and After

JOSEPH AND JESUS

(from the Spanish)

Said Joseph unto Mary,
  'Be counselled by me:
Fetch your love child from the manger,
  For to Egypt we must flee.'

As Mary went a-riding
  Up the hill out of view,
The ass was much astonished
  How like a dove he flew.

Said Jesus unto Joseph,
  Who his soft cheek did kiss:
'There are thorns in your beard, good sir.
  I askèd not for this.'

Then Joseph brought to Jesus
  Hot paps of white bread
Which, when it burned that pretty mouth,
  Joseph swallowed in his stead.

Robert Graves

CRYING, MY LITTLE ONE?

Crying, my little one, footsore and weary?
  Fall asleep, pretty one, warm on my shoulder:
I must tramp on through the winter night dreary,
  While the snow falls on me colder and colder.

You are my one, and I have not another;
  Sleep soft, my darling, my trouble and treasure;
Sleep warm and soft in the arms of your mother,
  Dreaming of pretty things, dreaming of pleasure.

Christina Rossetti

9
Epilogue

A CHRISTMAS BLESSING

God bless the master of this house,
  The mistress also,
And all the little children
That round the table go;
And all your kin and kinsfolk,
  That dwell both far and near:
I wish you a Merry Christmas
  And a Happy New Year.
                    Anon.

Index of First Lines

INDEX OF FIRST LINES

Index of Authors

Acknowledgments

The editor and publisher wish to thank the following for permission to reprint copyright material included in this anthology:

the *Daily Mirror* for 'Christmas to me' by *Thomas Boyle*;

the poet and the publisher (represented by David Higham Associates) for 'Innocent's Song' by *Charles Causley* from JOHNNY ALLELIUA published by Rupert Hart-Davis;

Miss D. E. Collins (represented by A. P. Watt & Son) for 'How Far is it to Bethlehem?' by *Frances Chesterton*;

Miss D. E. Collins and Methuen (represented by A. P. Watt & Son) for 'A Christmas Carol' by *G. K. Chesterton* from THE COLLECTED POEMS OF G.K. CHESTERTON;

the publisher for 'Singing in the Streets' and 'Bells Ringing' from SINGING IN THE STREETS by *Leonard Clark*, published by Dennis Dobson;

the Literary Trustees of Walter de la Mare, and the Society of Authors as their representative, for 'Now all the roads', 'Winter', and 'Dim-berried is the mistletoe' by *Walter de la Mare*;

Faber and Faber Ltd for 'Journey of the Magi' by *T. S. Eliot* from COLLECTED POEMS 1909–1962;

the Estate of Eleanor Farjeon and the publisher (represented by David Higham Associates) for 'Here we come again', 'When trees did show no leaves', 'Now every child', 'God bless your house' and 'In a Far Land' by *Eleanor Farjeon* from SILVER SAND AND SNOW published by Michael Joseph;

the Estate of Kenneth Grahame (represented by Curtis Brown Ltd, London) and Charles Scribner's Sons, New York, for 'Carol of the Field Mice' ('Villagers all, this frosty tide') from THE WIND IN THE WILLOWS by *Kenneth Grahame* (1908);

the poet (represented by A. P. Watt & Son) for 'Joseph and Jesus' from ANN AT HIGHWOOD HALL by *Robert Graves*;

Mrs George Bambridge and the publishers (represented by A. P. Watt & Son) for 'A Carol' and 'Eddi's Service' by *Rudyard Kipling* from REWARDS AND FAIRIES published by Macmillan & Co.;

the poet for 'The Eve of Christmas' by *James Kirkup*;

the Executors of the Estate of C. Day Lewis and the publisher, together with the Hogarth Press, for 'The Christmas Tree' from COLLECTED POEMS 1954 by *C. Day Lewis* published by Jonathan Cape Ltd;

the Society of Authors and the poet for 'Sing, happy child' by *Eiluned Lewis*;

ACKNOWLEDGMENTS

the Edinburgh University Press for 'The Computer's First Christmas Card' by *Edwin Morgan;*

the poet and the publisher (represented by David Higham Associates) for 'The Innkeeper's Wife' from THE WITNESSES by *Clive Sansom*, published by Methuen;

the poet for 'The Mayor and the Simpleton' by *Ian Serraillier;*

Mrs I. Wise, The Macmillan Company of Canada and Macmillan, London and Basingstoke, for 'A Singing in the Air' from 'Christmas at Freelands' by *James Stephens;*

Macmillan, London and Basingstoke, for 'Lullaby' and 'Afterthought' from THE SECRET BROTHER by *Elizabeth Jennings;*

the Trustees for the Copyrights of the late Dylan Thomas, and the publishers, for the extract from QUITE EARLY ONE MORNING by *Dylan Thomas*, published by J. M. Dent & Sons Ltd;

the publishers for 'Holly red and mistletoe white' from LITTLE GREY RABBIT'S CHRISTMAS by *Alison Uttley*, published by Collins.